BUS

CHEMICAL AND PROCESS DESIGN HANDBOOK

CHEMICAL AND PROCESS DESIGN HANDBOOK

James G. Speight

McGraw-Hill

New York   Chicago   San Francisco   Lisbon   London   Madrid
Mexico City   Milan   New Delhi   San Juan   Seoul
Singapore   Sydney   Toronto

Library of Congress Cataloging-in-Publication Data

Speight, J. G.
    Chemical and process design handbook / James Speight.
        p.   cm.
    Includes index.
    ISBN 0-07-137433-7 (acid-free paper)
    1. Chemical processes.   I. Title.
TP155.7 .S63   2002
660'.2812—dc21   2001052555

McGraw-Hill

A Division of The McGraw-Hill Companies

1 2 3 4 5 6 7 8 9 0   DOC/DOC   0 9 8 7 6 5 4 3 2 1

ISBN 0-07-137433-7

*The sponsoring editor for this book was Kenneth P. McCombs, the editing super-
visor was David E. Fogarty, and the production supervisor was Pamela A.
Pelton. It was set in the HB1A design in Times Roman by Kim Sheran, Deirdre
Sheean, and Vicki Hunt of McGraw-Hill Professional's Hightstown, New Jersey,
composition unit.*

Printed and bound by R. R. Donnelley & Sons Company.

This book was printed on recycled, acid-free paper containing
a minimum of 50% recycled, de-inked fiber.

CONTENTS

v

PREFACE

Chemicals are part of our everyday lives. The hundreds of chemicals that are manufactured by industrial processes influence what we do and how we do it. This book offers descriptions and process details of the most popular of those chemicals. The manufacture of chemicals involves many facets of chemistry and engineering which are exhaustively treated in a whole series of encyclopedic works, but it is not always simple to rapidly grasp present status of knowledge from these sources. Thus, there is a growing demand for a text that contains concise descriptions of the most important chemical conversions and processes of industrial operations.

This text will, therefore, emphasize the broad principles of systems of chemicals manufacture rather than intimate and encyclopedic details that are often difficult to understand. As such, the book will allow the reader to appreciate the chemistry and engineering aspects of important precursors and intermediates as well as to follow the development of manufacturing processes to current state-of-the-art processing.

This book emphasizes *chemical conversions*, which may be defined as chemical reactions applied to industrial processing. The basic chemistry will be set forth along with easy-to-understand descriptions, since the nature of the *chemical reaction* will be emphasized in order to assist in the understanding of reactor type and design. An outline is presented of the production of a range of chemicals from starting materials into useful products. These chemical products are used both as consumer goods and as intermediates for further chemical and physical modification to yield consumer products.

Since the basis of chemical-conversion classification is a chemical one, emphasis is placed on the important industrial *chemical reactions* and *chemical processes* in Part 1 of this book. These chapters focus on the various chemical reactions and the type of equipment that might be used in such processes. The contents of this part are in alphabetical order by reaction name.

Part 2 presents the reactions and processes by which individual chemicals, or chemical types, are manufactured and is subdivided by alphabetical listing of the various chemicals. Each item shows the chemical reaction by which that particular chemical can be manufactured. Equations are kept simple so that they can be understood by people in the many scientific and engineering disciplines involved in the chemical manufacturing industry. Indeed, it is hoped that the chemistry is sufficiently simple that nontechnical readers can understand the equations.

The design of equipment can often be simplified by the generalizations arising from a like chemical-conversion arrangement rather than by considering each reaction as unique.

Extensive use of *flowcharts* is made as a means of illustrating the various processes and to show the main reactors and the paths of the feedstocks and products. However, no effort is made to include all of the valves and ancillary equipment that might appear in a true industrial setting. Thus, the flowcharts used here have been reduced to maximum simplicity and are designed to show principles rather than details.

Although all chemical manufacturers should be familiar with the current selling prices of the principal chemicals with which they are concerned, providing price information is not a purpose of this book. Prices per unit weight or volume are subject to immediate changes and can be very misleading. For such information, the reader is urged to consult the many sources that deal with the prices of chemical raw materials and products.

In the preparation of this work, the following sources have been used to provide valuable information:

AIChE Journal (AIChE J.)

Canadian Journal of Chemistry

Canadian Journal of Chemical Engineering

Chemical and Engineering News (Chem. Eng. News)

ChemTech

Chemical Week (Chem. Week)

Chemical Engineering Progress (Chem. Eng. Prog.)

Chemical Processing Handbook, J. J. McKetta (ed.), Marcel Dekker, New York.

Encyclopedia of Chemical Technology, 4th ed., It. E. Kirk, and D. F. Othmer(eds.) Wiley-Interscience, New York

Chemical Engineers' Handbook, 7th ed., R. H. Perry and D. W. Green (eds.), McGraw-Hill, New York.

Chemical Processing

Handbook of Chemistry and Physics, Chemical Rubber Co.

Hydrocarbon Processing

Industrial and Engineering Chemistry (*Ind. Eng. Chem.*)

Industrial and Engineering Chemistry Fundamentals (*Ind. Eng. Chem. Fundamentals*)

Industrial and Engineering Chemistry Process Design and Development (*Ind. Eng. Chem. Process Des. Dev.*)

Industrial and Engineering Chemistry Product Research and Development (*Ind. Eng. Chem. Prod. Res. Dev.*)

International Chemical Engineering

Journal of Chemical and Engineering Data (*J. Chem. Eng. Data*)

Journal of the Chemical Society

Journal of the American Chemical Society

Lange's Handbook of Chemistry, 12th ed., J. A. Dean (ed.). McGraw-Hill, New York

Oil & Gas Journal

McGraw-Hill Encyclopedia of Science and Technology, 5th ed., McGraw-Hill, New York

Riegel's Industrial Chemistry, 7th ed., J. A. Kent (ed.), Reinhold, New York

Finally, I am indebted to my colleagues in many different countries who have continued to engage me in lively discussions and who have offered many thought-provoking comments about industrial processes. Such contacts were of great assistance in the writing of this book and have been helpful in formulating its contents.

James G. Speight

Part 1
REACTION TYPES

ALKYLATION

Alkylation is usually used to increase performance of a product and involves the conversion of, for example, an amine to its alkylated homologs as in the reaction of aniline with methyl alcohol in the presence of sulfuric acid catalyst:

$$C_6H_5NH_2 + 2CH_3OH \rightarrow C_6H_5N(CH_3)_2 + 2H_2O$$

Thus, aniline, with a considerable excess of methyl alcohol and a catalytic amount of sulfuric acid, is heated in an autoclave at about 200°C for 5 or 6 hours at a high reaction pressure of 540 psi (3.7 MPa). Vacuum distillation is used for purification.

In the alkylation of aniline to diethylaniline by heating aniline and ethyl alcohol, sulfuric acid cannot be used because it will form ether; consequently, hydrochloric acid is employed, but these conditions are so corrosive that the steel used to resist the pressure must be fitted with replaceable enameled liners.

Alkylation reactions employing alkyl halides are carried out in an acidic medium. For example, hydrobromic acid is formed when methyl bromide is used in the alkylation leading, and for such reactions an autoclave with a replaceable enameled liner and a lead-coated cover is suitable.

In the petroleum refining industry, alkylation is the union of an olefin with an aromatic or paraffinic hydrocarbon:

$$CH_2=CH_2 + (CH_3)_3CH \rightarrow (CH_3)_3CCH_2CH_3$$

Alkylation processes are exothermic and are fundamentally similar to refining industry polymerization processes but they differ in that only part of the charging stock need be unsaturated. As a result, the *alkylate* product contains no olefins and has a higher octane rating. These methods are based on the reactivity of the tertiary carbon of the *iso*-butane with olefins, such as propylene, butylenes, and amylenes. The product *alkylate* is a mixture of saturated, stable isoparaffins distilling in the gasoline range, which becomes a most desirable component of many high-octane gasolines.

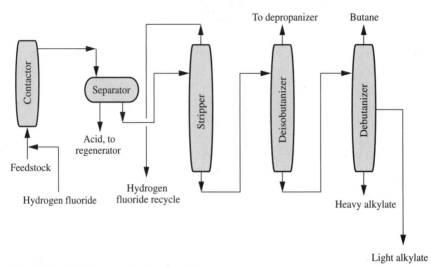

FIGURE 1   Alkylation using hydrogen fluoride.

Alkylation is accomplished by using either of two catalysts: (1) hydrogen fluoride and (2) sulfuric acid. In the alkylation process using liquid hydrogen fluoride (Fig. 1), the acid can be used repeatedly, and there is virtually no acid-disposal problem. The acid/hydrocarbon ratio in the contactor is 2:1 and temperature ranges from 15 to 35°C can be maintained since no refrigeration is necessary. The anhydrous hydrofluoric acid is regenerated by distillation with sufficient pressure to maintain the reactants in the liquid phase.

In many cases, steel is suitable for the construction of alkylating equipment, even in the presence of the strong acid catalysts, as their corrosive effect is greatly lessened by the formation of esters as catalytic intermediate products.

In the petroleum industry, the sulfuric acid and hydrogen fluoride employed as alkylation catalysts must be substantially anhydrous to be effective, and steel equipment is satisfactory. Where conditions are not anhydrous, lead-lined, monel-lined, or enamel-lined equipment is satisfactory. In a few cases, copper or tinned copper is still used, for example, in the manufacture of pharmaceutical and photographic products to lessen contamination with metals.

Distillation is usually the most convenient procedure for product recovery, even in those instances in which the boiling points are rather close together. Frequently such a distillation will furnish a finished material of quality sufficient to meet the demands of the market. If not, other means of purification may be necessary, such as crystallization or separation by means of solvents. The choice of a proper solvent will, in many instances, lead to the crystallization of the alkylated product and to its convenient recovery.

The converse reactions *dealkylation* and *hydrodealkylation* are practiced extensively to convert available feedstocks into other more desirable (marketable), products. Two such processes are: (1) the conversion of toluene or xylene, or the higher-molecular-weight alkyl aromatic compounds, to benzene in the presence of hydrogen and a suitable presence of a dealkylation catalyst and (2) the conversion of toluene in the presence of hydrogen and a fixed bed catalyst to benzene plus mixed xylenes.

AMINATION

Amination is the process of introducing the amino group ($-NH_2$) into an organic compound as, for example, the production of aniline ($C_6H_5NH_2$) by the reduction of nitrobenzene ($C_6H_5NO_2$) in the liquid phase (Fig. 1) or in the vapor phase in a fluidized bed reactor (Fig. 2). For many decades, the only method of putting an amino group on an aryl nucleus involved adding a nitro ($-NO_2$) group, then *reduction* to the amino ($-NH_2$) group.

Without high-pressure vessels and catalysts, reduction had to be done by reagents that would function under atmospheric pressure. The common reducing agents available under these restrictions are:

1. Iron and acid

2. Zinc and alkali

3. Sodium sulfide or polysulfide

4. Sodium hydrosulfite

5. Electrolytic hydrogen

6. Metal hydrides

Now liquid- and gas-phase hydrogenations can be performed on a variety of materials.

$$RNO_2 + 3H_2 \rightarrow RNH_2 + 2H_2O$$

Where metals are used to produce the reducing hydrogen, several difficult processing problems are created. The expense is so great that it is necessary to find some use for the reacted material. Spent iron can sometimes be used for pigment preparations or to absorb hydrogen sulfide. Stirring a vessel containing much metal is quite difficult.

On a small scale, cracking ammonia can produce hydrogen for reduction. Transport and storage of hydrogen as ammonia is compact, and the cracking procedure involves only a hot pipe packed with catalyst and

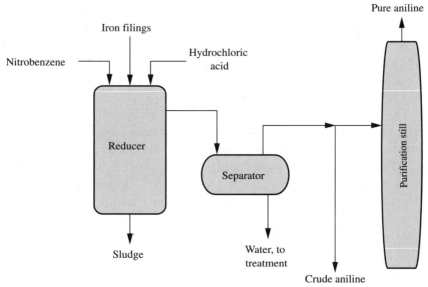

FIGURE 1   Aniline production by the reduction of nitrobenzene.

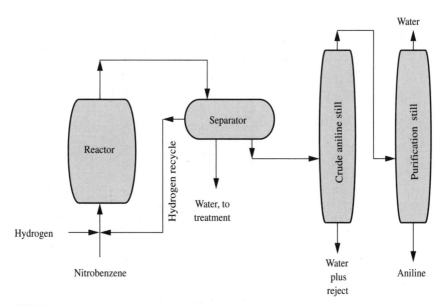

FIGURE 2   Vapor phase reduction of nitrobenzene to aniline.

immersed in a molten salt bath. The nitrogen that accompanies the generated hydrogen is inert.

Amination is also achieved by the use of ammonia (NH_3), in a process referred to as *ammonolysis*. An example is the production of aniline ($C_6H_5NH_2$) from chlorobenzene (C_6H_5Cl) with ammonia (NH_3). The reaction proceeds only under high pressure.

The replacement of a nuclear substituent such as hydroxyl (–OH), chloro, (–Cl), or sulfonic acid (–SO$_3$H) with amino (–NH$_2$) by the use of ammonia (*ammonolysis*) has been practiced for some time with feedstocks that have reaction-inducing groups present thereby making replacement easier. For example, 1,4-dichloro-2-nitrobenzene can be changed readily to 4-chloro-2-nitroaniline by treatment with aqueous ammonia. Other molecules offer more processing difficulty, and pressure vessels are required for the production of aniline from chlorobenzene or from phenol (Fig. 3).

$$C_6H_5OH + NH_3 \rightarrow C_6H_5NH_2 + H_2O$$

Ammonia is a comparatively low cost reagent, and the process can be balanced to produce the desired amine. The other routes to amines

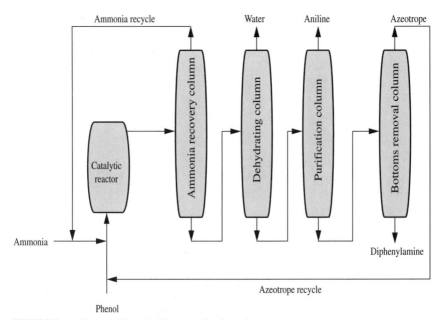

FIGURE 3   Aniline and diphenylamine production from phenol.

through reduction use expensive reagents (iron, Fe, zinc, Zn, or hydrogen, H_2, gas) that make ammonolysis costs quite attractive. Substituted amines can be produced by using substituted ammonia (amines) in place of simple ammonia. The equipment is an agitated iron pressure vessel; stainless steel is also used for vessel construction.

Amination by reduction is usually carried out in cast-iron vessels (1600 gallons capacity, or higher) and alkali reductions in carbon steel vessels of desired sizes. The vessel is usually equipped with a nozzle at the base so that the iron oxide sludge or entire charge may be run out upon completion of the reaction.

In some reducers, a vertical shaft carries a set of cast-iron stirrers to keep the iron particles in suspension in the lower part of the vessel and to maintain all the components of the reaction in intimate contact. In addition, the stirrer assists in the diffusion of the amino compound away from the surface of the metal and thereby makes possible a more extensive contact between nitro body and catalytic surface.

Thus, amination, or reaction with ammonia, is used to form both aliphatic and aromatic amines. Reduction of nitro compounds is the traditional process for producing amines, but ammonia or substituted ammonias (amines) react directly to form amines. The production of aniline by amination now exceeds that produced by reduction (of nitrobenzene).

Oxygen-function compounds also may be subjected to ammonolysis, for example:

1. Methanol plus aluminum phosphate catalyst yields monomethylamine (CH_3NH_2), dimethylamine [$(CH_3)2NH$], and trimethylamine [$(CH_3)3N$]

2. 2-naphthol plus sodium ammonium sulfite ($NaNH_3SO_3$) catalyst (Bucherer reaction) yields 2-naphthylamine

3. Ethylene oxide yields monoethanolamine ($HOCH_2CH_2NH_2$), diethanolamine [$(HOCH_2CH_2)_2NH)$], and triethanolamine [$(HOCH_2CH_2)_3N)$]

4. Glucose plus nickel catalyst yields glucamine

5. Cyclohexanone plus nickel catalyst yields cyclohexylamine

Methylamines are produced by reacting gaseous methanol with a catalyst at 350 to 400°C and 290 psi (2.0 MPa), then distilling the reaction mixture. Any ratio of mono-, di-, or trimethylamines is possible by recycling the unwanted products.

An equilibrium mixture of the three ethanolamines is produced when ethylene oxide is bubbled through 28% aqueous ammonia at 30 to 40°C. By recirculating the products of the reaction, altering the temperatures, pressures, and the ratio of ammonia to ethylene oxide, but always having an excess of ammonia, it is possible to make the desired amine predominate. Diluent gas also alters the product ratio.

$$CH_2CH_2O + NH_3 \rightarrow HOCH_2CH_2NH_2 + H_2O$$
monoethanolamine

$$2CH_2CH_2O + NH_3 \rightarrow (HOCH_2CH_2)_2NH + 2H_2O$$
diethanolamine

$$3CH_2CH_2O + NH_3 \rightarrow (HOCH_2CH_2)_3N + 3H_2O$$
triethanolamine

After the strongly exothermic reaction, the reaction products are recovered and separated by flashing off and recycling the ammonia, and then fractionating the amine products.

Monomethylamine is used in explosives, insecticides, and surfactants. Dimethylamine is used for the manufacture of dimethylformamide and acetamide, pesticides, and water treatment. Trimethylamine is used to form choline chloride and to make biocides and slimicides.

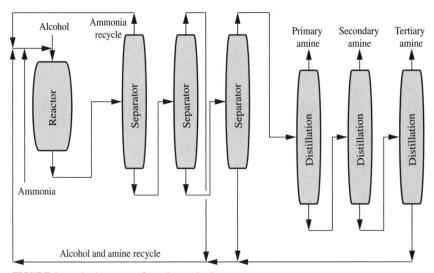

FIGURE 4   Amination process for amine production.

Other alkylamines can be made in similar fashion from the alcohol and ammonia (Fig. 4). Methyl, ethyl, isopropyl, cyclohexyl, and combination amines have comparatively small markets and are usually made by reacting the correct alcohol with anhydrous ammonia in the vapor phase.

CONDENSATION
AND ADDITION

There are only a few products manufactured in any considerable tonnage by condensation and addition (Friedel-Crafts) reactions, but those that are find use in several different intermediates and particularly in making high-quality vat dyes.

The agent employed in this reaction is usually an acid chloride or anhydride, catalyzed with aluminum chloride. Phthalic anhydride reacts with chlorobenzene to give p-chlorobenzoylbenzoic acid and, in a continuing action, the p-chlorobenzoylbenzoic acid forms β-chloroanthraquinone.

Since anthraquinone is a relatively rare and expensive component of coal tar and petroleum, this type of reaction has been the basis for making relatively inexpensive anthraquinone derivatives for use in making many fast dyes for cotton.

Friedel-Crafts reactions are highly corrosive, and the aluminum-containing residues are difficult to dispose.

DEHYDRATION

Dehydration is the removal of water or the elements of water, in the correct proportion, from a substance or system or chemical compound. The elements of water may be removed from a single molecule or from more than one molecule, as in the dehydration of alcohol, which may yield ethylene by loss of the elements of water from one molecule or ethyl ether by loss of the elements of water from two molecules:

$$CH_3CH_2OH \rightarrow CH_2{=}CH_2 + H_2O$$

$$2CH_3CH_2OH \rightarrow CH_3CH_2OCH_2CH_3 + H_2O$$

The latter reaction is commonly used in the production of ethers by the dehydration of alcohols.

Vapor-phase dehydration over catalysts such as alumina is also practiced. Hydration of olefins to produce alcohols, usually over an acidic catalyst, produces substantial quantities of ethers as by-products. The reverse reaction, ethers to alcohols, can be accomplished by recycling the ethers over a catalyst.

In food processing, dehydration is the removal of more than 95% of the water by use of thermal energy. However, there is no clearly defined line of demarcation between *drying* and *dehydrating*, the latter sometimes being considered as a supplement of drying.

The term *dehydration* is not generally applied to situations where there is a loss of water as the result of evaporation. The distinction between the terms drying and dehydrating may be somewhat clarified by the fact that most substances can be dried beyond their capability of restoration.

Rehydration or *reconstitution* is the restoration of a dehydrated food product to its original edible condition by the simple addition of water, usually just prior to consumption or further processing.

DEHYDROGENATION

Dehydrogenation is a reaction that results in the removal of hydrogen from an organic compound or compounds, as in the dehydrogenation of ethane to ethylene:

$$CH_3CH_3 \rightarrow CH_2{=}CH_2 + H_2$$

This process is brought about in several ways. The most common method is to heat hydrocarbons to high temperature, as in thermal cracking, that causes some dehydrogenation, indicated by the presence of unsaturated compounds and free hydrogen.

In the chemical process industries, nickel, cobalt, platinum, palladium, and mixtures containing potassium, chromium, copper, aluminum, and other metals are used in very large-scale dehydrogenation processes.

Styrene is produced from ethylbenzene by dehydrogenation (Fig. 1). Many lower molecular weight aliphatic *ketones* are made by *dehydration*

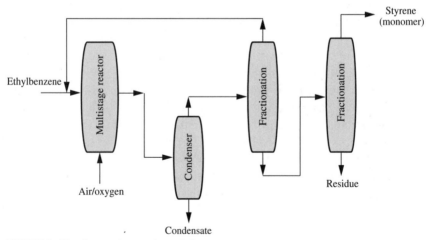

FIGURE 1   Manufacture of styrene from ethylbenzene.

of alcohols. Acetone, methyl ethyl ketone, and cyclohexanone can be made in this fashion.

$$C_6H_5CH_2CH_3 \rightarrow C_6H_5CH{=}CH_2 + H_2$$

Acetone is the ketone used in largest quantity and is produced as a by-product of the manufacture of phenol via cumene. Manufacture from *iso*-propanol is by the reaction:

$$(CH_3)_2CHOH \rightarrow (CH_3)_2C{=}O$$

This reaction takes place at 350°C and 200 kPa with copper or zinc acetate as the catalyst; conversion is 85 to 90 percent. Purification by distillation follows.

The dehydrogenation of *n*-paraffins yields detergent alkylates and *n*-olefins. The catalytic use of rhenium for selective dehydrogenation has increased in recent years since dehydrogenation is one of the most commonly practiced of the chemical unit processes.

*See **Hydrogenation.***

ESTERIFICATION

A variety of solvents, monomers, medicines, perfumes, and explosives are made from esters of nitric acid. Ethyl acetate, n-butyl acetate, iso-butyl acetate, glycerol trinitrate, pentaerythritol tetranitrate (PETN), glycol dinitrate, and cellulose nitrate are examples of such reactions.

Ester manufacture is a relatively simple process in which the alcohol and an acid are heated together in the presence of a sulfuric acid catalyst, and the reaction is driven to completion by removing the products as formed (usually by distillation) and employing an excess of one of the reagents. In the case of ethyl acetate, esterification takes place in a column that takes a ternary azeotrope. Alcohol can be added to the condensed overhead liquid to wash out the alcohol, which is then purified by distillation and returned to the column to react.

Amyl, butyl, and iso-propyl acetates are all made from acetic acid and the appropriate alcohols. All are useful lacquer solvents and their slow rate of evaporation (compared to acetone or ethyl acetate) prevents the surface of the drying lacquer from falling below the dew point, which would cause condensation on the film and a mottled surface appearance (*blushing*). Other esters of importance are used in perfumery and in plasticizers and include methyl salicylate, methyl anthranilate, diethyl-phthalate, dibutyl-phthalate, and di-2-ethylhexyl-phthalate.

Unsaturated vinyl esters for use in polymerization reactions are made by the esterification of olefins. The most important ones are vinyl esters: vinyl acetate, vinyl chloride, acrylonitrile, and vinyl fluoride. The addition reaction may be carried out in either the liquid, vapor, or mixed phases, depending on the properties of the acid. Care must be taken to reduce the polymerization of the vinyl ester produced.

Esters of allyl alcohol, e.g., diallyl phthalate, are used as bifunctional polymerization monomers and can be prepared by simple esterification of phthalic anhydride with allyl alcohol. Several acrylic esters, such as ethyl or methyl acrylates, are also widely used and can be made from acrylic acid and the appropriate alcohol. The esters are more volatile than the corresponding acids.

ETHYNYLATION

The *ethynylation* reaction involves the addition of acetylene to carbonyl compounds.

$$HC{\equiv}CH + R^1COR^2 \rightarrow HC{\equiv}CC(OH)R^1R^2$$

Heavy metal acetylides, particularly cuprous acetylide ($CuC{\equiv}CH$), catalyze the addition of acetylene ($HC{\equiv}CH$) to aldehydes ($RCH{=}O$).

FERMENTATION

Fermentation processes produce a wide range of chemicals that complement the various chemicals produced by nonfermentation routes. For example, alcohol, acetone, butyl alcohol, and acetic acid are produced by fermentation as well as by synthetic routes. Almost all the major antibiotics are obtained from fermentation processes.

Fermentation under controlled conditions involves chemical conversions, and some of the more important processes are:

1. *Oxidation,* e.g., ethyl alcohol to acetic acid, sucrose to citric acid, and dextrose to gluconic acid

2. *Reduction,* e.g., aldehydes to alcohols (acetaldehyde to ethyl alcohol) and sulfur to hydrogen sulfide

3. *Hydrolysis,* e.g., starch to glucose and sucrose to glucose and fructose and on to alcohol

4. E*sterification,* e.g., hexose phosphate from hexose and phosphoric acid

FRIEDEL-CRAFTS REACTIONS

Several chemicals are manufactured by application of the Friedel-Crafts condensation reaction. Efficient operation of any such process depends on:

1. The preparation and handling of reactants
2. The design and construction of the apparatus
3. The control of the reaction so as to lead practically exclusively to the formation of the specific products desired
4. The storage of the catalyst (aluminum chloride)

Several of the starting reactants, such as acid anhydrides, acid chlorides, and alkyl halides, are susceptible to hydrolysis. The absorption of moisture by these chemicals results in the production of compounds that are less active, require more aluminum chloride for condensation, and generally lead to lower yields of desired product. Furthermore, the ingress of moisture into storage containers for these active components usually results in corrosion problems.

Anhydrous aluminum chloride needs to be stored in iron drums under conditions that ensure the absence of moisture. When, however, moisture contacts the aluminum chloride, hydrogen chloride is formed, the quantity of hydrogen chloride thus formed depends on the amount of water and the degree of agitation of the halide. If sufficient moisture is present, particularly in the free space in the container or reaction vessel or at the point of contact with the outside atmosphere, then hydrochloric acid is formed and leads to corrosion of the storage container.

In certain reactions, such as the isomerization of butane and the alkylation of isoparaffins, problems of handling hydrogen chloride and acidic sludge are encountered. The corrosive action of the aluminum chloride–hydrocarbon complex, particularly at 70 to 100°C, has long been recognized and various reactor liners have been found satisfactory.

The rate of reaction is a function of the efficiency of the contact between the reactants, i.e., stirring mechanism and mixing of the reactants. In fact, mixing efficiency has a vital influence on the yield and purity of the product. Insufficient or inefficient mixing may lead to uncondensed reactants or to excessive reaction on heated surfaces.

HALOGENATION

Halogenation is almost always chlorination, for the difference in cost between chlorine and the other halogens, particularly on a molar basis, is quite substantial. In some cases, the presence of bromine (Br), iodine (I), or fluorine (F) confers additional properties to warrant manufacture.

Chlorination proceeds (1) by addition to an unsaturated bond, (2) by substitution for hydrogen, or (3) by replacement of another group such as hydroxyl (–OH) or sulfonic (–SO$_3$H). Light catalyzes some chlorination reactions, temperature has a profound effect, and polychlorination almost always occurs to some degree. All halogenation reactions are strongly exothermic.

In the chlorination process (Fig.1), chlorine and methane (fresh and recycled) are charged in the ratio 0.6/1.0 to a reactor in which the temperature is maintained at 340 to 370°C. The reaction product contains chlorinated hydrocarbons with unreacted methane, hydrogen chloride, chlorine, and heavier chlorinated products. Secondary chlorination reactions take place at ambient temperature in a light-catalyzed reactor that converts methylene chloride to chloroform, and in a reactor that converts chloroform to carbon tetrachloride. By changing reagent ratios, temperatures, and recycling ratio, it is possible to vary the product mix somewhat to satisfy market demands. Ignition is avoided by using narrow channels and high velocities in the reactor. The chlorine conversion is total, and the methane conversion around 65 percent.

Equipment for the commercial chlorination reactions is more difficult to select, since the combination of halogen, oxygen, halogen acid, water, and heat is particularly corrosive. Alloys such as Hastelloy and Durichlor resist well and are often used, and glass, glass-enameled steel, and tantalum are totally resistant but not always available. Anhydrous conditions permit operation with steel or nickel alloys. With nonaqueous media, apparatus constructed of iron and lined with plastics and/or lead and glazed tile is the most suitable, though chemical stoneware, fused quartz, glass, or glass-lined equipment can be used for either the whole plant or specific apparatus.

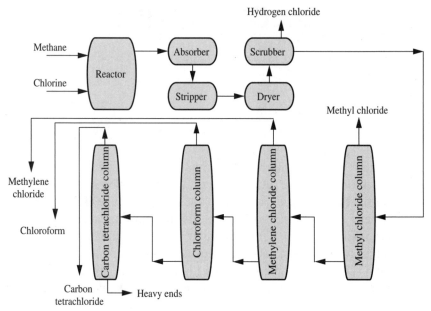

FIGURE 1   Production of chloromethanes by chlorination of methane.

When chlorination has to be carried out at a low temperature, it is often beneficial to circulate cooling water through a lead coil within the chlorinator or circulate the charge through an outside cooling system rather than to make use of an external jacket. When the temperature is to be maintained at 0°C or below, a calcium chloride brine, cooled by a refrigerating machine, is employed.

Most chlorination reactions produce hydrogen chloride as a by-product, and a method was searched for to make this useful for further use:

$$4HCl + O_2 \rightarrow 2H_2O + 2Cl_2$$

However, this is not a true equilibrium reaction, with a tendency to favor hydrogen chloride. The reaction can be used and driven to completion by use of the *oxychlorination* procedure that reacts the chlorine with a reactive substance as soon as it is formed, thus driving the reaction to completion as, for example, in the oxychlorination of methane:

$$CH_4 + HCl + O_2 \rightarrow CH_3Cl + CH_2Cl_2 + CHCl_3 + CCl_4 + H_2O$$

This chlorination can be accomplished with chlorine but a mole of hydrogen chloride is produced for every chlorine atom introduced into the methane, and this must be disposed of to prevent environmental pollution. Thus, the use of by-product hydrogen chloride from other processes is frequently available and the use of cuprous chloride (CuCl) and cupric chloride (CuCl$_2$), along with some potassium chloride (KCl) as a molten salt catalyst, enhances the reaction progress.

Ethane can be chlorinated under conditions very similar to those for methane to yield mixed chlorinated ethanes.

Chlorobenzene is used as a solvent and for the manufacture of nitrochlorobenzenes. It is manufactured by passing dry chlorine through benzene, using ferric chloride (FeCl$_3$) as a catalyst:

$$C_6H_6 + Cl_2 \rightarrow C_6H_5Cl + HCl$$

The reaction rates favor production of chlorobenzene over dichloroben-zene by 8.5:1, provided that the temperature is maintained below 60°C. The hydrogen chloride generated is washed free of chlorine with benzene, then absorbed in water. Distillation separates the chlorobenzene, leaving mixed isomers of dichlorobenzene.

In aqueous media, when hydrochloric acid is present in either the liquid or vapor phase and particularly when under pressure, tantalum is undoubt-edly the most resistant material of construction. Reactors and catalytic tubes lined with this metal give satisfactory service for prolonged periods.

HYDRATION AND HYDROLYSIS

Ethyl alcohol is a product of fermentation of sugars and cellulose but the alcohol is manufactured mostly by the hydration of ethylene.

An indirect process for the manufacture of ethyl alcohol involves the dissolution of ethylene in sulfuric acid to form ethyl sulfate, which is hydrolyzed to form ethyl alcohol (Fig. 1). There is always some by-product diethyl ether that can be either sold or recirculated.

$$3CH_2=CH_2 + 2H_2SO_4 \rightarrow C_2H_5HSO4 + (C_2H_5)_2SO_4$$

$$C_2H_5HSO_4 + (C_2H_5)_2SO_4 + H_2O \rightarrow 3C_2H_5OH + 2H_2SO_4$$

$$C_2H_5OH + C_2H_5HSO_4 \rightarrow C_2H_5OC_2H_5$$

The conversion yield of ethylene to ethyl alcohol is 90 percent with a 5 to 10 percent yield of diethyl ether $(C_2H_5OC_2H_5)$.

A direct hydration method using phosphoric acid as a catalyst at 300°C is also available (Fig. 2):

$$CH_2=CH_2 + H_2O \rightarrow C_2H_5OH$$

and produces ethyl alcohol in yields in excess of 92 percent. The conversion per pass is 4 to 25 percent, depending on the activity of the catalyst used.

In this process, ethylene and water are combined with a recycle stream in the ratio ethylene/water 1/0.6 (mole ratio), a furnace heats the mixture to 300°C, and the gases react over the catalyst of phosphoric acid absorbed on diatomaceous earth. Unreacted reagents are separated and recirculated. By-product acetaldehyde (CH_3CHO) is hydrogenated over a catalyst to form more ethyl alcohol.

Iso-propyl alcohol is a widely used and easily made alcohol. It is used in making acetone, cosmetics, chemical derivatives, and as a process solvent. There are four processes that are available for the manufacture of *iso*-propyl alcohol:

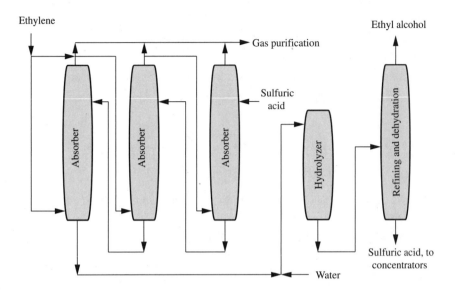

FIGURE 1 Manufacture of ethyl alcohol from ethylene and sulfuric acid.

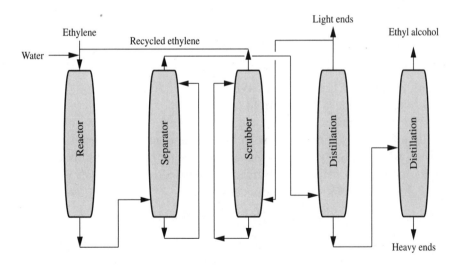

FIGURE 2 Manufacture of ethyl alcohol by direct hydration.

1. A sulfuric acid process similar to the one described for ethanol hydration
2. A gas-phase hydration using a fixed-bed-supported phosphoric acid catalyst
3. A mixed-phase reaction using a cation exchange resin catalyst
4. A liquid-phase hydration in the presence of a dissolved tungsten catalyst

The last three processes (2, 3, and 4) are all essentially direct hydration processes.

$$CH_3CH=CH_2 + H_2O \rightarrow CH_3CHOHCH_3$$

Per-pass conversions vary from a low of 5 to a high of 70 percent for the gas-phase reaction.

Secondary butanol ($CH_3CH_2CHOHCH_3$) is manufactured by processes similar to those described for ethylene and propylene.

Hydrolysis usually refers to the replacement of a sulfonic group ($-SO_3H$) or a chloro group ($-Cl$) with an hydroxyl group ($-OH$) and is usually accomplished by fusion with alkali. Hydrolysis uses a far wider range of reagents and operating conditions than most chemical conversion processes.

Polysubstituted molecules may be hydrolyzed with less drastic conditions. Enzymes, acids, or sometimes water can also bring about hydrolysis alone.

$$ArSO_3Na + 2NaOH \rightarrow ArONa + Na_2SO_3 + H_2O$$

$$ArCl + 2NaOH \rightarrow ArONa + NaCl + H_2O$$

Acidification will give the hydroxyl compound (ArOH). Most hydrolysis reactions are modestly exothermic.

The more efficient route via cumene has superceded the fusion of benzene sulfonic acid with caustic soda for the manufacture of phenol, and the hydrolysis of chlorobenzene to phenol requires far more drastic conditions and is no longer competitive. Ethylene chlorohydrin can be hydrolyzed to glycol with aqueous sodium carbonate.

$$ClCH_2CH_2OH \rightarrow HOCH_2CH_2OH$$

Cast-iron or steel open fusion pots heated to the high temperatures required (200 to 325°C) with oil, electricity, or directly with gas, are standard equipment.

HYDROFORMYLATION

The hydroformylation (oxo) reactions offer ways of converting a-olefins to aldehydes and/or alcohols containing an additional carbon atom.

$$CH_3CH{=}CH_2 + CO + H_2 \rightarrow CH_3CH_2CH_2CHO$$

$$CH_3CH_2CH_2CHO + H_2 \rightarrow CH_3CH_2CH_2CH_2OH$$

In the process (Fig. 1), the olefin in a liquid state is reacted at 27 to 30 MPa and 150 to 170°C in the presence of a soluble cobalt catalyst. The aldehyde and a lesser amount of the alcohol are formed and flashed off along with steam, and the catalyst is recycled. Conversions of over 97 percent are obtained, and the reaction is strongly exothermic. The carbon monoxide and hydrogen are usually in the form of *synthesis gas*.

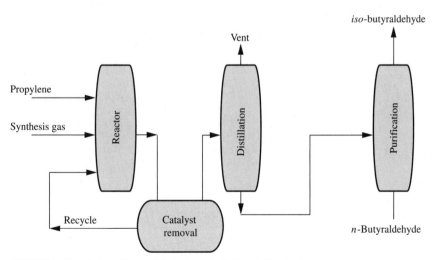

FIGURE 1 Manufacture of butyraldehyde by the hydroformylation (oxo) reaction.

When propylene is used as the hydrocarbon, *n*- and *iso*-butyraldehyde are formed. This reaction is most frequently run with the C_3 and C_7 to C_{12} olefins. When C_7 olefins are used, a series of dimethyl- and ethylhexanols and methyl heptanols are formed that are used as octyl alcohols to make plasticizers and esters.

*See **Oxo Reaction.***

HYDROGENATION

In its simplest interpretation, hydrogenation is the addition of hydrogen to a chemical compound. Generally, the process involves elevated temperature and relatively high pressure in the presence of a catalyst.

Hydrogenation yields many useful chemicals, and its use has increased phenomenally, particularly in the petroleum refining industry. Besides saturating double bonds, hydrogenation can be used to eliminate other elements from a molecule. These elements include oxygen, nitrogen, halogens, and particularly sulfur. Cracking (thermal decomposition) in the presence of hydrogen is particularly effective in desulfurizing high-boiling petroleum fractions, thereby producing lower-boiling and higher-quality products.

Although occasionally hydrogen for a reaction is provided by donor solvents and a few older reactions use hydrogen generated by acid or alkali acting upon a metal, gaseous hydrogen is the usual hydrogenating agent.

Hydrogenation is generally carried out in the presence of a catalyst and under elevated temperature and pressure. Noble metals, nickel, copper, and various metal oxide combinations are the common catalysts.

Nickel, prepared in finely divided form by reduction of nickel oxide in a stream of hydrogen gas at about 300°C, was introduced by 1897 as a catalyst for the reaction of hydrogen with unsaturated organic substances to be conducted at about 175°C. Nickel proved to be one of the most successful catalysts for such reactions. The unsaturated organic substances that are hydrogenated are usually those containing a double bond, but those containing a triple bond also may be hydrogenated. Platinum black, palladium black, copper metal, copper oxide, nickel oxide, aluminum, and other materials have subsequently been developed as hydrogenation catalysts. Temperatures and pressures have been increased in many instances to improve yields of desired product. The hydrogenation of methyl ester to fatty alcohol and methanol, for example, occurs at about 290 to 315°C and 3000 psi (20.7 MPa). In the hydrotreating of liquid hydrocarbon fuels to improve quality, the reaction may take place in fixed-bed reactors at pressures ranging from 100 to 3000 psi (690 kPa to 20.7 MPa).

Many hydrogenation processes are of a proprietary nature, with numerous combinations of catalysts, temperature, and pressure possible.

Lower pressures and higher temperatures favor dehydrogenation, but the catalysts used are the same as for hydrogenation.

Methyl alcohol (methanol) is manufactured from a mixture of carbon monoxide and hydrogen (synthesis gas), using a copper-based catalyst.

$$CO + 2H_2 \rightarrow CH_3OH$$

In the process (Fig. 1), the reactor temperature is 250 to 260°C at a pressure of 725 to 1150 psi (5 to 8 MPa). High- and low-boiling impurities are removed in two columns and the unreacted gas is recirculated.

New catalysts have helped increase the conversion and yields. The older, high-pressure processes used zinc-chromium catalysts, but the low-pressure units use highly active copper catalysts. Liquid-entrained micrometer-sized catalysts have been developed that can convert as much as 25 percent per pass. Contact of the synthesis gases with hot iron catalyzes competing reactions and also forms volatile iron carbonyl that fouls the copper catalyst. Some reactors are lined with copper.

Because the catalyst is sensitive to sulfur, the gases are purified by one of several sulfur-removing processes, then are fed through heat exchangers into one of two types of reactors. With bed-in-place reactors, steam at around 4.5 kPa, in quantity sufficient to drive the gas compressors, can be generated. A tray-type reactor with gases introduced just above every bed

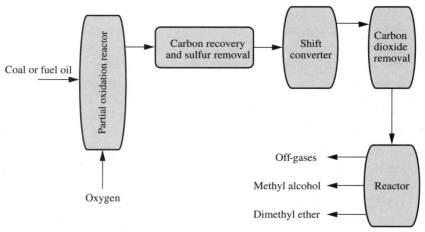

FIGURE 1   Manufacture of methyl alcohol from synthesis gas.

for cooling offers more nearly isothermal operation but does not give convenient heat recovery.

Reaction vessels are usually of two types: one in which the contents are agitated or stirred in some way and the other in which the reactor and contents are stationary. The first is used with materials such as solids or liquids that need to be brought into intimate contact with the catalyst and the hydrogen. The second type is used where the substance may have sufficient vapor pressure at the temperature of operation so that a gas-phase as well as a liquid-phase reaction is possible. It is also most frequently used in continuous operation where larger quantities of material need to be processed than can be done conveniently with batch methods.

In hydrogenation processes, heating of the ingoing materials is best accomplished by heat exchange with the outgoing materials and adding additional heat by means of high-pressure pipe coils. A pipe coil is the only convenient and efficient method of heating, for the reactor is usually so large that heating it is very difficult. It is usually better practice to add all the heat needed to the materials before they enter the reactor and then simply have the reactor properly insulated thermally. Hydrogenation reactions are usually exothermic, so that once the process is started, the problem may be one of heat removal. This is accomplished by allowing the heat of reaction to flow into the ingoing materials by heat exchange in the reactor, or, if it is still in excess, by recycling and cooling in heat exchangers the proper portion of the material to maintain the desired temperature.

See **Dehydrogenation.**

NITRATION

Nitration is the insertion of a nitro group ($-NO_2$) into an organic compound, usually through the agency of the reaction of a hydrocarbon with nitric acid. Concentrated sulfuric acid may be used as a catalyst.

$$ArH + HNO_3 \rightarrow ArNO_2 + H_2O$$

More than one hydrogen atom may be replaced, but replacement of each succeeding hydrogen atom represents a more difficult substitution.

The nitrogen-bearing reactant may be:

1. Strong nitric acid
2. Mixed nitric and sulfuric acid
3. A nitrate plus sulfuric acid
4. Nitrogen pentoxide (N_2O_5)
5. A nitrate plus acetic acid

Both straight chain and ring-type carbon compounds can be nitrated; alkanes yield nitroparaffins.

The process for the production of nitrobenzene from benzene involves the use of mixed acid (Fig. 1), but there are other useful nitrating agents, e.g., inorganic nitrates, oxides of nitrogen, nitric acid plus acetic anhydride, and nitric acid plus phosphoric acid. In fact, the presence of sulfuric acid in quantity is vital to the success of the nitration because it increases the solubility of the hydrocarbon in the reaction mix, thus speeding up the reaction, and promotes the ionization of the nitric acid to give the nitronium ion (NO_2^+), which is the nitrating species. Absorption of water by sulfuric acid favors the nitration reaction and shifts the reaction equilibrium to the product.

Nitration offers a method of making unreactive paraffins into reactive substances without cracking. Because nitric acid and nitrogen oxides are strong oxidizing agents, oxidation always accompanies nitration. Aromatic nitration reactions have been important particularly for the manufacture of explosives. Nitrobenzene is probably the most important nitration product.

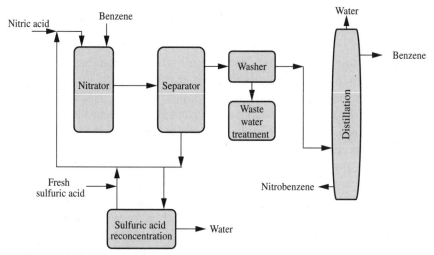

FIGURE 1   Production of nitrobenzene from benzene.

Certain esters of nitric acid (cellulose nitrate, glyceryl trinitrate) are often referred to as nitro compounds (nitrocellulose, nitroglycerin), but this terminology should be avoided.

Vapor-phase nitration of paraffin hydrocarbons, particularly propane, can be brought about by uncatalyzed contact between a large excess of hydrocarbon and nitric acid vapor at around 400°C, followed by quenching. A multiplicity of nitrated and oxidized products results from nitrating propane; nitromethane, nitroethane, nitropropanes, and carbon dioxide all appear, but yields of useful products are fair. Materials of construction must be very oxidation-resistant and are usually of ceramic-lined steel. The nitroparaffins have found limited use as fuels for race cars, submarines, and model airplanes. Their reduction products, the amines, and other hydroxyl compounds resulting from aldol condensations have made a great many new aliphatic syntheses possible because of their ready reactivity.

Nitration reactions are carried out in closed vessels that are provided with an agitating mechanism and means for controlling the reaction temperature. The nitration vessels are usually constructed of cast iron and steel, but often acid-resistant alloys, particularly chrome-nickel steel alloys, are used.

Plants may have large (several hundred gallon capacity) nitration vessels operating in a batch mode or small continuous units. The temperature is held at about 50°C, governed by the rate of feed of benzene. Reaction is rapid in well-stirred and continuous nitration vessels. The reaction products are decanted from the spent acid and are washed with dilute alkali. The spent acid is sent to some type of recovery system and yields of 98 percent can be anticipated.

Considerable heat evolution accompanies the nitration reaction, oxidation increases it, and the heat of dilution of the sulfuric acid increases it still further. Increased temperature favors dinitration arid oxidation, so the reaction must be cooled to keep it under control. Good heat transfer can be assured by the use of jackets, coils, and good agitation in the nitration vessel. Nitration vessels are usually made of stainless steel, although cast iron stands up well against mixed acid.

When temperature regulation is dependent solely on external jackets, a disproportional increase in nitration vessel capacity as compared with jacket surface occurs when the size of the machine is enlarged. Thus, if the volume is increased from 400 to 800 gallons, the heat-exchange area increases as the square and the volume as the cube of the expanded unit. To overcome this fault, internal cooling coils or tubes are introduced, which have proved satisfactory when installed on the basis of sound calculations that include the several thermal factors entering into this unit process.

A way of providing an efficient agitation inside the nitration vessel is essential if local overheating is to be mitigated. Furthermore, the smoothness of the reaction depends on the dispersion of the reacting material as it comes in contact with the change in the nitration vessel so that a fairly uniform temperature is maintained throughout the vessel.

Nitration vessels are usually equipped with one of three general types of agitating mechanism: (1) single or double impeller, (2) propeller or turbine, with cooling sleeve, and (3) outside tunnel circulation.

The *single-impeller* agitator consists of one vertical shaft containing horizontal arms. The shaft may be placed off center in order to create rapid circulation past, or local turbulence at, the point of contact between the nitrating acid and the organic compound.

The *double-impeller* agitator consists of two vertical shafts rotating in opposite directions, and each shaft has a series of horizontal arms attached. The lower blades have an upward thrust, whereas the upper ones repel the liquid downward. This conformation provides a reaction mix that is essentially homogeneous.

The term *sleeve-and-propeller* agitation is usually applied when the nitration vessel is equipped with a vertical sleeve through which the charge is circulated by the action of a marine propeller or turbine. The sleeve is usually made of a solid bank of acid-resisting cooling coils through which cold water or brine is circulated at a calculated rate. In order to obtain the maximum efficiency with this type of nitration vessel, it is essential to maintain a rapid circulation of liquid upward or downward in the sleeves and past the coils.

OXIDATION

Oxidation is the addition of oxygen to an organic compound or, conversely, the removal of hydrogen.

Reaction control is the major issue during oxidation reactions. Only partial oxidation is required for conversion of one organic compound into another or complete oxidation to carbon dioxide and water will ensue.

The most common oxidation agent is air, but oxygen is frequently used. Chemical oxidizing agents (nitric acid, dichromates, permanganates, chromic anhydride, chlorates, and hydrogen peroxide) are also often used.

As examples of oxidation processes, two processes are available for the manufacture of phenol, and both involve oxidation. The major process involves oxidation of cumene to cumene hydroperoxide, followed by decomposition to phenol and acetone. A small amount of phenol is also made by the oxidation of toluene to benzoic acid, followed by decomposition of the benzoic acid to phenol.

Benzoic acid is synthesized by liquid-phase toluene oxidation over a cobalt naphthenate catalyst with air as the oxidizing agent. An older process involving halogenation of toluene to benzotrichloride and its decomposition into benzoic acid is still used available.

Maleic acid and anhydride are recovered as by-products of the oxidation of xylenes and naphthalenes to form phthalic acids, and are also made specifically by the partial oxidation of benzene over a vanadium pentoxide (V_2O_5) catalyst. This is a highly exothermic reaction, and several modifications of the basic process exist, including one using butylenes as the starting materials.

Formic acid is made by the oxidation of formamide or by the liquid-phase oxidation of n-butane to acetic acid. The by-product source is expected to dry up in the future, and the most promising route to replace it is through carbonylation of methanol.

Caprolactam, adipic acid, and hexamethylenediamine (HMDA) are all made from cyclohexane. Almost all high-purity cyclohexane is obtained by hydrogenating benzene, although some for solvent use is obtained by careful distillation of selected petroleum fractions.

Several oxidative routes are available to change cyclohexane to cyclo-hexanone, cyclohexanol, and ultimately to adipic acid or caprolactam. If phenol is hydrogenated, cyclohexanone can be obtained directly; this will react with hydroxylamine to give cyclohexanone oxime that converts to caprolactam on acid rearrangement. Cyclohexane can also be converted to adipic acid, then adiponitrile, which can be converted to hexamethylenedi-amine. Adipic acid and hexamethylenediamine are used to form nylon 6,6. This route to hexamethylenediamine is competitive with alternative routes through butene.

Acetaldehyde is manufactured by one of several possible processes: (1) the hydration of acetylene, no longer a significant process. (2) the Wacker pro-cess, in which ethylene is oxidized directly with air or 99% oxygen (Fig. 1) in the presence of a catalyst such as palladium chloride with a copper chloride promoter. The ethylene gas is bubbled, at atmospheric pressure, through the solution at its boiling point. The heat of reaction is removed by boiling of the water. Unreacted gas is recycled following condensation of the aldehyde and water, which are then separated by distillation, (3) passing ethyl alco-hol over a copper or silver gauze catalyst gives a 25 percent conversion to acetaldehyde, with recirculation making a 90 to 95 percent yield possible, and (4) a process in which lower molecular weight paraffin hydrocarbons are oxidized noncatalytically to produce mixed compounds, among them acetaldehyde and acetic acid.

Liquid-phase reactions in which oxidation is secured by the use of oxi-dizing compounds need no special apparatus in the sense of elaborate means for temperature control and heat removal. There is usually provided a kettle form of apparatus, closed to prevent the loss of volatile materials and fitted with a reflux condenser to return vaporized materials to the reac-

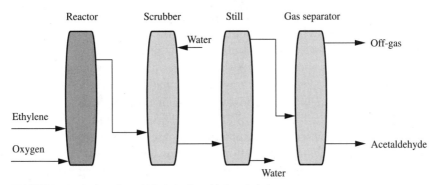

FIGURE 1 Production of acetaldehyde by the oxidation of ethylene.

tion zone, with suitable means for adding reactants rapidly or slowly as may be required and for removing the product, and provided with adequate jackets or coils through which heating or cooling means may be circulated as required.

In the case of liquid-phase reactions in which oxidation is secured by means of atmospheric oxygen—for example, the oxidation of liquid hydrocarbons to fatty acids—special means must be provided to secure adequate mixing and contact of the two immiscible phases of gaseous oxidizing agent and the liquid being oxidized. Although temperature must be controlled and heat removed, the requirements are not severe, since the temperatures are generally low and the rate of heat generation controllable by regulation of the rate of air admission.

Heat may be removed and temperature controlled by circulation of either the liquid being oxidized or a special cooling fluid through the reaction zone and then through an external heat exchanger. Mixing may be obtained by the use of special distributor inlets for the air, designed to spread the air throughout the liquid and constructed of materials capable of withstanding temperatures that may be considerably higher at these inlet ports than in the main body of the liquid. With materials that are sensitive to overoxidation and under conditions where good contact must be used partly to offset the retarding effect of necessarily low temperatures, thorough mixing may be provided by the use of mechanical stirring or frothing of the liquid.

By their very nature, the vapor-phase oxidation processes result in the concentration of reaction heat in the catalyst zone, from which it must be removed in large quantities at high-temperature levels. Removal of heat is essential to prevent destruction of apparatus, catalyst, or raw material, and maintenance of temperature at the proper level is necessary to ensure the correct rate and degree of oxidation. With plant-scale operation and with reactions involving deep-seated oxidation, removal of heat constitutes a major problem. With limited oxidation, however, it may become necessary to supply heat even to oxidations conducted on a plant scale.

In the case of vapor-phase oxidation of aliphatic substances such as methanol and the lower molecular weight aliphatic hydrocarbons, the ratio of reacting oxygen is generally lower than in the case of the aromatic hydrocarbons for the formation of the desired products, and for this reason heat removal is simpler. Furthermore, in the case of the hydrocarbons, the proportion of oxygen in the reaction mixture is generally low, resulting in low per-pass conversions and, in some instances, necessitating preliminary heating of the reactants to reaction temperature.

Equipment for the oxidation of the aromatic hydrocarbons requires that the reactor design permit the maintenance of elevated temperatures, allow the removal of large quantities of heat at these elevated temperatures, and provide adequate catalyst surface to promote the reactions.

OXO REACTION

The *oxo reaction* is the general or generic name for a process in which an unsaturated hydrocarbon is reacted with carbon monoxide and hydrogen to form oxygen function compounds, such as aldehydes and alcohols.

In a typical process for the production of oxo alcohols, the feedstock comprises an olefin stream, carbon monoxide, and hydrogen. In a first step, the olefin reacts with CO and H2 in the presence of a catalyst (often cobalt) to produce an aldehyde that has one more carbon atom than the originating olefin:

$$RCH=CH_2 + CO + H_2 \rightarrow RCH_2CH_2CH=O$$

This step is exothermic and requires an ancillary cooling operation.

The raw aldehyde exiting from the oxo reactor then is subjected to a higher temperature to convert the catalyst to a form for easy separation from the reaction products. The subsequent treatment also decomposes unwanted by-products. The raw aldehyde then is hydrogenated in the presence of a catalyst (usually nickel) to form the desired alcohol:

$$RCH_2CH_2CH=O + H_2 \rightarrow RCH_2CH_2CH_2OH$$

The raw alcohol then is purified in a fractionating column. In addition to the purified alcohol, by-products include a light hydrocarbon stream and a heavy oil. The hydrogenation step takes place at about 150°C under a pressure of about 1470 psi (10.13 MPa). The olefin conversion usually is about 95 percent.

Among important products manufactured in this manner are substituted propionaldehyde from corresponding substituted ethylene, normal and *iso*-butyraldehyde from propylene, *iso*-octyl alcohol from heptene, and trimethylhexyl alcohol from di-*iso*butylene.

See **Hydroformylation**.

POLYMERIZATION

Polymerization is a process in which similar molecules (usually olefins) are linked to form a high-molecular-weight product; such as the formation of polyethylene from ethylene

$$nCH_2CH_2 \rightarrow H-(CH_2CH_2)_n-H$$

The molecular weight of the polyethylene can range from a few thousand to several hundred thousand.

Polymerization of the monomer in bulk may be carried out in the liquid or vapor state. The monomers and activator are mixed in a reactor and heated or cooled as needed. As most polymerization reactions are exothermic, provision must be made to remove the excess heat. In some cases, the polymers are soluble in their liquid monomers, causing the viscosity of the solution to increase greatly. In other cases, the polymer is not soluble in the monomer and it precipitates out after a small amount of polymerization occurs.

In the petroleum industry, the term *polymerization* takes on a different meaning since the polymerization processes convert by-product hydrocarbon gases produced in cracking into liquid hydrocarbons suitable (of limited or specific molecular weight) for use as high-octane motor and aviation fuels and for petrochemicals.

To combine olefinic gases by polymerization to form heavier fractions, the combining fractions must be unsaturated. Hydrocarbon gases, particularly olefins, from cracking reactors are the major feedstock of polymerization.

$$(CH_3)_2C=CH_2 \rightarrow (CH_3)_3CH_2C(CH_3)=CH_2$$
$$(CH_3)_3CH_2C(CH_3)=CH_2 \rightarrow C_{12}H_{24}$$

Vapor-phase cracking produces considerable quantities of unsaturated gases suitable as feedstocks for polymerization units.

Catalytic polymerization is practical on both large and small scales and is adaptable to combination with reforming to increase the quality of the gasoline. Gasoline produced by polymerization contains a smog-producing olefinic bond. Polymer oligomers are widely used to make detergents.

SULFONATION

Sulfonation is the introduction of a sulfonic acid group ($-SO_3H$) into an organic compound as, for example, in the production of an aromatic sulfonic acid from the corresponding aromatic hydrocarbon.

$$ArH + H_2SO_4 \rightarrow ArSO_3H + H_2O$$

The usual sulfonating agent is concentrated sulfuric acid, but sulfur trioxide, chlorosulfonic acid, metallic sulfates, and sulfamic acid are also occasionally used. However, because of the nature and properties of sulfuric acid, it is desirable to use it for nucleophilic substitution wherever possible.

For each substance being sulfonated, there is a critical concentration of acid below which sulfonation ceases. The removal of the water formed in the reaction is therefore essential. The use of a very large excess of acid, while expensive, can maintain an essentially constant concentration as the reaction progresses. It is not easy to volatilize water from concentrated solutions of sulfuric acid, but azeotropic distillation can sometimes help.

The sulfonation reaction is exothermic, but not highly corrosive, so sulfonation can be conducted in steel, stainless-steel, or cast-iron sulfonators. A jacket heated with hot oil or steam can serve to heat the contents sufficiently to get the reaction started, then carry away the heat of reaction. A good agitator, a condenser, and a fume control system are usually also provided.

1- and 2-naphthalenesulfonic acids are formed simultaneously when naphthalene is sulfonated with concentrated sulfuric acid. The isomers must be separated if pure α- or β-naphthol are to be prepared from the product mix. Variations in time, temperature, sulfuric acid concentration, and acid/hydrocarbon ratio alter the yields to favor one particular isomer, but a pure single substance is never formed. Using similar acid/hydrocarbon ratios, sulfonation at 40°C yields 96% alpha isomer, 4% beta, while at 160°C the proportions are 15% α-naphthol, 8.5% β-naphthol.

The α-sulfonic acid can be hydrolyzed to naphthalene by passing steam at 160°C into the sulfonation mass. The naphthalene so formed passes out with the steam and can be recovered. The pure β-sulfonic acid left behind can be hydrolyzed by caustic fusion to yield relatively pure β-naphthol.

In general, separations are based on some of the following consideration:

1. Variations in the *rate* of hydrolysis of two isomers
2. Variations in the solubility of various salts in water
3. Differences in solubility in solvents other than water
4. Differences in solubility accentuated by common-ion effect (salt additions)
5. Differences in properties of derivatives
6. Differences based on molecular size, such as using molecular sieves or absorption.

Sulfonation reactions may be carried out in batch reactors or in continuous reactors. Continuous sulfonation reactions are feasible only when the organic compounds possess certain chemical and physical properties, and are practical in only a relatively few industrial processes. Most commercial sulfonation reactions are batch operations.

Continuous operations are feasible and practical (1) where the organic compound (benzene or naphthalene) can be volatilized, (2) when reaction rates are high (as in the chlorosulfonation of paraffins and the sulfonation of alcohols), and (3) where production is large (as in the manufacture of detergents, such as alkylaryl sulfonates).

Water of reaction forms during most sulfonation reactions, and unless a method is devised to prevent excessive dilution because of water formed during the reaction, the rate of sulfonation will be reduced. In the interests of economy in sulfuric acid consumption, it is advantageous to remove or chemically combine this water of reaction. For example, the use of reduced pressure for removing the water of reaction has some technical advantages in the sulfonation of phenol and of benzene.

The use of the partial-pressure distillation is predicated upon the ability of the diluent, or an excess of volatile reactant, to remove the water of reaction as it is formed and, hence, to maintain a high concentration of sulfuric acid. If this concentration is maintained, the necessity for using excess sulfuric acid is eliminated, since its only function is to maintain the acid concentration above the desired value. Azeotropic removal of the water of reaction in the sulfonation of benzene can be achieved by using an excess of vaporized benzene.

The use of oleum ($H_2SO_4 \cdot SO_3$) for maintaining the necessary sulfur trioxide concentration of a sulfonation mixture is a practical procedure. Preferably the oleum and organic compound should be added gradually and concurrently to a large volume of cycle acid so as to take up the water as rapidly as it is formed by the reaction. Sulfur trioxide may be added intermittently to the sulfonation reactor to maintain the sulfur trioxide concentration above the value for the desired degree of sulfonation.

VINYLATION

Unlike ethynylation, in which acetylene adds across a carbonyl group and the triple bond is retained, in vinylation a labile hydrogen compound adds to acetylene, forming a double bond.

$$XH + HC\equiv CH \rightarrow CH_2=CHX$$

Catalytic vinylation has been applied to the manufacture of a wide range of alcohols, phenols, thiols, carboxylic acids, and certain amines and amides. Vinyl acetate is no longer prepared this way in many countries, although some minor vinyl esters such as vinyl stearate may still be manufactured by this route. However, the manufacture of vinyl-pyrrolidinone and vinyl ethers still depends on acetylene as the starting material.

Part 2

MANUFACTURE OF CHEMICALS

ACETALDEHYDE

Acetaldehyde (ethanal, $CH_3CH=O$, melting point $-123.5°C$, boiling point: $20.1°C$, density: 0.7780, flash point: $-38°C$, ignition temperature: $165°C$) is a colorless, odorous liquid.

Acetaldehyde has a pungent, suffocating odor that is somewhat fruity and quite pleasant in dilute concentrations. Acetaldehyde is miscible in all proportions with water and most common organic solvents, e.g., acetone, benzene, ethyl alcohol, ether, gasoline, toluene, xylenes, turpentine, and acetic acid.

Because of its versatile chemical reactivity, acetaldehyde is widely used as a commencing material in organic syntheses, including the production of resins, dyestuffs, and explosives. It is also used as a reducing agent, preservative, and medium for silvering mirrors. In resin manufacture, paraldehyde $[(CH_3CHO)_3]$ sometimes is preferred because of its higher boiling and flash points.

Acetaldehyde was first prepared by Scheele in 1774, by the action of manganese dioxide (MnO_2) and sulfuric acid (H_2SO_4) on ethyl alcohol (ethanol, CH_3CH_2OH).

$$CH_3CH_2OH + [O] \rightarrow CH_3CH=O + H_2O$$

Commercially, passing alcohol vapors and preheated air over a silver catalyst at 480°C carries out the oxidation. With a multitubular reactor, conversions of 74 to 82 percent per pass can be obtained while generating steam to be used elsewhere in the process.

The formation of acetaldehyde by the addition of water to acetylene was observed by Kutscherow in 1881.

$$HC≡CH + H_2O \rightarrow CH_3CH=O$$

In this hydration process, high-purity acetylene under a pressure of 15 psi (103.4 kPa) is passed into a vertical reactor containing a mercury catalyst dissolved in 18 to 25% sulfuric acid at 70 to 90°C. Fresh catalyst is fed to the reactor periodically; the catalyst may be added in the mercurous (Hg^+)

form, but the catalytic species has been shown to be a mercuric ion complex. The excess acetylene sweeps out the dissolved acetaldehyde, which is condensed by water and refrigerated brine and then scrubbed with water; this crude acetaldehyde is purified by distillation; the unreacted acetylene is recycled. The catalytic mercuric ion is reduced to catalytically inactive mercurous sulfate (Hg_2SO_4) and metallic mercury. Sludge, consisting of reduced catalyst and tars, is drained from the reactor at intervals and resulfated. The rate of catalyst depletion can be reduced by adding ferric or other suitable ions to the reaction solution. These ions reoxidize the mercurous ion to the mercuric ion; consequently, the quantity of sludge that must be recovered is reduced.

In one variation of the process, acetylene is completely hydrated with water in a single operation at 68 to 73°C using the mercuric-iron salt catalyst. The acetaldehyde is partially removed by vacuum distillation and the mother liquor recycled to the reactor. The aldehyde vapors are cooled to about 35°C, compressed to 37 psi (253 kPa), and condensed. It is claimed that this combination of vacuum and pressure operations substantially reduces heating and refrigeration costs.

The commercial process of choice for acetaldehyde production is the direct oxidation of ethylene.

$$CH_2{=}CH_2 + [O] \rightarrow CH_3CH{=}O$$

There are two variations for this commercial production route: the two-stage process and the one-stage process.

In the one-stage process (Fig. 1), ethylene, oxygen, and recycle gas are directed to a vertical reactor for contact with the catalyst solution under

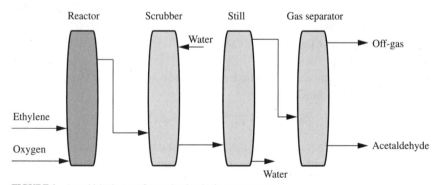

FIGURE 1   Acetaldehyde manufacture by the single-stage process.

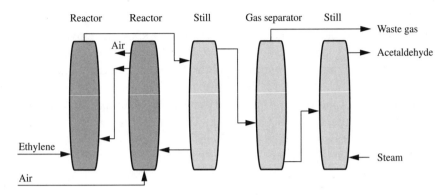

FIGURE 2   Acetaldehyde manufacture by the two-stage process.

slight pressure. The water evaporated during the reaction absorbs the heat evolved, and makeup water is fed as necessary to maintain the desired catalyst concentration. The gases are water scrubbed, and the resulting acetaldehyde solution is fed to a distillation column. The tail gas from the scrubber is recycled to the reactor. Inert materials are eliminated from the recycle gas in a bleed stream that flows to an auxiliary reactor for additional ethylene conversion.

In the two-stage process (Fig. 2), ethylene is almost completely oxidized by air to acetaldehyde in one pass in a tubular plug-flow reactor made of titanium. The reaction is conducted at 125 to 130°C and 150 psi (1.03 MPa) with the palladium and cupric chloride catalysts. Acetaldehyde produced in the first reactor is removed from the reaction loop by adiabatic flashing in a tower. The flash step also removes the heat of reaction. The catalyst solution is recycled from the flash-tower base to the second stage (or oxidation reactor), where the cuprous salt is oxidized to the cupric state with air. The high-pressure off-gas from the oxidation reactor, mostly nitrogen, is separated from the liquid catalyst solution and scrubbed to remove acetaldehyde before venting. A small portion of the catalyst stream is heated in the catalyst regenerator to destroy any undesirable copper oxalate. The flasher overhead is fed to a distillation system where water is removed for recycle to the reactor system and organic impurities, including chlorinated aldehydes, are separated from the purified acetaldehyde product. Synthesis techniques purported to reduce the quantity of chlorinated by-products generated have been patented.

Acetaldehyde was first used extensively during World War I as a starting material for making acetone (CH_3COCH_3) from acetic acid

(CH_3CO_2H) and is currently an important intermediate in the production of acetic acid, acetic anhydride $(CH_3CO-O-OCCH_3)$, ethyl acetate $(CH_3CO-OC_2H_5)$, peracetic acid $(CH_3CO-O-OH)$ and a variety of other chemicals such as pentaerythritol, chloral, glyoxal, alkylamines, and pyridines.

In aqueous solutions, acetaldehyde exists in equilibrium with the acetaldehyde hydrate $[CH_3CH(OH)_2]$. The enol form, vinyl alcohol $(CH_2=CHOH)$ exists in equilibrium with acetaldehyde to the extent of 0.003% (1 molecule in approximately 30,000) and can be acetylated with ketene $(CH_2=C=O)$ to form vinyl acetate $(CH_2=CHOCOCH_3)$.

ACETAL RESINS

Acetal resins are those homopolymers (melting point: ca. 175°C, density: ca. 1.41) and copolymers (melting point: ca. 165°C, density: ca. 1.42) where the backbone or main structural chain is completely or essentially composed of repeating oxymethylene units $(-CH_2O-)_n$. The polymers are derived chiefly from formaldehyde (methanal, $CH_2=O$), either directly or through its cyclic trimer, trioxane or 1,3,5-trioxacyclohexane.

Formaldehyde polymerizes by both anionic and cationic mechanisms. Strong acids are needed to initiate cationic polymerization and anionic polymerization is initiated by relatively weak bases (e.g., pyridine). Boron trifluoride (BF_3) or other Lewis acids are used to promote polymerization where trioxane is the raw material.

In the process, anhydrous formaldehyde is continuously fed to a reactor containing well-agitated inert solvent, especially a hydrocarbon, in which monomer is sparingly soluble. Initiator, especially amine, and chain-transfer agent are also fed to the reactor. The reaction is quite exothermic and polymerization temperature is maintained below 75°C (typically near 40°C) by evaporation of the solvent. The product polymer is not soluble in the solvent and precipitates early in the reaction.

The polymer is separated from the polymerization slurry and slurried with acetic anhydride and sodium acetate catalyst. Acetylation of polymer end groups is carried out in a series of stirred tank reactors at temperatures up to 140°C. End-capped polymer is separated by filtration and washed at least twice, once with acetone and then with water.

The copolymerization of trioxane with cyclic ethers or formals is accomplished with cationic initiators such as boron trifluoride dibutyl etherate. Polymerization by ring opening of the six-membered ring to form high molecular weight polymer does not commence immediately upon mixing monomer and initiator. Usually, an induction period is observed during which an equilibrium concentration of formaldehyde is produced.

When the equilibrium formaldehyde concentration is reached, the polymer begins to precipitate and further polymerization takes place in trioxane solution, and more comonomer is exhausted at relatively low conversion, but a random copolymer is nevertheless obtained.

In the process, molten trioxane, initiator, and comonomer are fed to the reactor; a chain-transfer agent is included if desired. Polymerization proceeds in bulk with precipitation of polymer, and the reactor must supply enough shearing to continually break up the polymer bed, reduce particle size, and provide good heat transfer. Raw copolymer is obtained as fine crumb or flake containing imbibed formaldehyde and trioxane that are substantially removed in subsequent treatments which may be combined with removal of unstable end groups.

Acetal copolymer may be end capped in a process completely analogous to that used for homopolymer. However, the presence of comonomer units (e.g., -O-CH$_2$-CH$_2$-O-) in the backbone and the relative instability to base of hemiacetal end groups allow for another convenient route to a polymer with stable end groups. The hemiacetal end groups may be subjected to base-catalyzed (especially amine) hydrolysis in the melt or in solution or suspension, and the chain segments between the end group and the nearest comonomer unit deliberately depolymerized until the depropagating chain encounters the comonomer unit. If ethylene oxide or dioxolane is used as comonomer, a stable hydroxyethyl ether end group results (-O-CH$_2$CH$_2$-OH). Some formate end groups, which are intermediate in thermal stability between hemiacetal and ether end groups, may also be removed by this process.

The product from the melt or suspension treatment is obtained directly as crumb or powder. The polymer recovered from solution treatment is obtained by precipitative cooling or spray drying. The polymer with now stable end groups may be washed and dried to remove impurities, especially acids or their precursors, prior to finishing operations.

The average molecular weight MW of acetal copolymers may be estimated from their melt index (MI, expressed in g/10 min):

$$MI = 3.3 \times 10^{18}MW^{-3.55}$$

Stiffness, resistance to deformation under constant applied load (creep resistance), resistance to damage by cyclical loading (fatigue resistance), and excellent lubricity are mechanical properties for which acetal resins are perhaps best known and which have contributed significantly to their excellent commercial success. General-purpose acetal resins are substantially stiffer than general-purpose polyamides (nylon-6 or -6,6 types) when the latter have reached equilibrium water content.

Acetal resins are generally stable in mildly alkaline environments. However, bases can catalyze hydrolysis of ester end groups, resulting in a less thermally stable polymer.

Acetals provide excellent resistance to most organic compounds except when exposed for long periods at elevated temperatures. The resins have limited resistance to strong acids and oxidizing agents. The copolymers and some of the homopolymers are resistant to the action of weak bases. Normally, where resistance to burning, weathering, and radiation are required, acetals are not specified. The resins are used for cams, gears, bearings, springs, sprockets, and other mechanical parts, as well as for electrical parts, housings, and hardware.

ACETAMINOPHEN

Acetaminophen, sold under the trade name Tylenol, is a widely used analgesic and antipyretic that is an over-the-counter drug. Combined with codeine it is one of the top five prescription drugs. Acetaminophen is prepared by treating *p*-aminophenol with a mixture of glacial acetic acid and acetic anhydride.

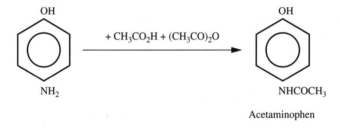

Acetaminophen

ACETIC ACID

Acetic acid (ethanoic acid, vinegar acid, CH_3CO_2H, melting point 16.6°C, boiling point: 117.9°C, density: 1.0490, flash point: 43°C, ignition temperature 465°C) is a colorless, pungent liquid that is miscible with water, alcohol, and ether in all proportions.

Acetic acid is available commercially in several concentrations: (1) glacial acetic is approximately 99.7% glacial acetic acid with water the principal impurity, (2) reagent grade acetic acid generally contains 36% acetic acid by weight, and (3) commercial aqueous solutions are usually 28, 56, 70, 80, 85, and 90% acetic acid.

Acetic acid is the active ingredient in vinegar, in which the content ranges from 4 to 5% acetic acid. Acetic acid is classified as a weak, monobasic acid ($-CO_2H$) but the three hydrogen atoms linked to the carbon atom (CH_3) are not replaceable by metals.

Acetic acid is manufactured by three processes: acetaldehyde oxidation, n-butane oxidation, and methanol carbonylation.Ethylene is the exclusive organic raw material for making acetaldehyde, 70 percent of which is further oxidized to acetic acid or acetic anhydride. The single-stage (Wacker) process for making acetaldehyde involves cupric chloride and a small amount of palladium chloride in aqueous solution as a catalyst.

$$CH_2=CH_2 + H_2O + PdCl_2 \rightarrow CH_3CHO + 2HCl + Pd^0$$

The yield is 95 percent and further oxidation of the acetaldehyde produces acetic acid (Fig. 1).

$$2CH_3CHO + O_2 \rightarrow 2CH_3CO_2H$$

A manganese or cobalt acetate catalyst is used with air as the oxidizing agent in the temperature and pressure ranges of 55 to 80°C and 15 to 75 psi; the yield is 95 percent.

The second manufacturing method for acetic acid utilizes butane from the C_4 petroleum stream rather than ethylene. A variety of products is formed but conditions can be controlled to allow a large percentage of acetic acid to be formed. Cobalt, manganese, or chromium acetates are cat-

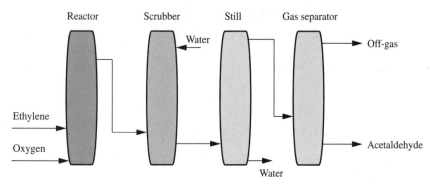

FIGURE 1   Acetaldehyde manufacture by the single-stage (Wacker) process.

alysts with temperatures of 50 to $-250^\circ C$ and a pressure of 800 psi.

$$C_4H_{10} + O_2 \rightarrow CH_3CO_2H + HCO_2H + CH_3CH_2OH + CH_3OH$$

The third and preferred method of acetic acid manufacture is the carbonylation of methanol, involving reaction of methanol and carbon monoxide (both derived from methane) with rhodium and iodine as catalysts at $175^\circ C$ and 1 atm (Fig. 2).

$$CH_3OH + CO \rightarrow CH_3CO_2H$$

The yield of acetic acid is 99 percent based on methanol and 90 percent based on carbon monoxide.

Acetic acid is used for the manufacture of methyl acetate (Fig. 3) and acetic anhydride (Fig. 4), vinyl acetate, ethyl acetate, terephthalic acid, cellulose acetate, and a variety of acetic esters.

Vinyl acetate is used mainly as a fiber in clothing. Ethyl acetate is a com-

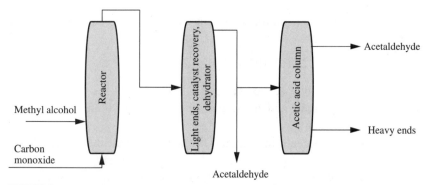

FIGURE 2   Acetaldehyde manufacture by carbonylation of methyl alcohol (methanol).

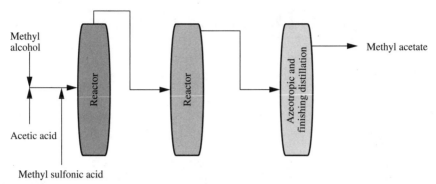

FIGURE 3 Methyl acetate manufacture from methyl alcohol (methanol).

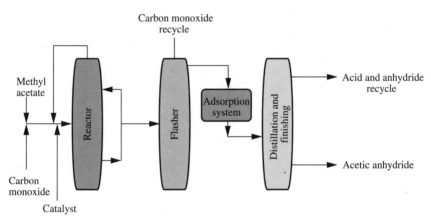

FIGURE 4 Acetic anhydride manufacture by carbonylation of from methyl acetate.

mon organic solvent. Acetic acid is used in the manufacture of terephthalic acid, which is a monomer for the synthesis of poly (ethylene terephthalate), the *polyester* of the textile industry.

ACETIC ANHYDRIDE

Acetic anhydride (boiling point: 139.5, density: 1.0820) may be produced by three different methods. The first procedure involves the in situ production from acetaldehyde of peracetic acid, which in turn reacts with more acetaldehyde to yield the anhydride.

$$CH_3CH{=}O + O_2 \rightarrow CH_3C({=}O)OH$$

$$CH_3C({=}O)OH + CH_3CH{=}O \rightarrow CH_3C({=}O)O(O{=})C\,CH_3 + H_2O$$

In the preferred process, acetic acid (or acetone) is pyrolyzed to ketene, which reacts with acetic acid to form acetic anhydride.

$$CH_3C({=}O)OH \rightarrow CH_2{=}C{=}O + H_2O$$

$$CH_2{=}C{=}O + CH_3C({=}O)OH \rightarrow CH_3C({=}O)O(O{=})C\,CH_3$$

Another process to make acetic anhydride involves carbon monoxide insertion into methyl acetate (Fig. 1).

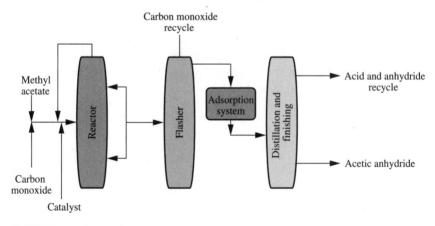

FIGURE 1   Acetic anhydride manufacture by carbonylation of methyl acetate

$$CH_3C(=O)OCH_3 \rightarrow CH_3C(=O)O(O=)CCH_3$$

Approximately 80 percent of acetic anhydride is used as a raw material in the manufacture of cellulose acetate.

ACETONE

Acetone (dimethyl ketone, 2-propanone, CH_3COCH_3, melting point: −94.6°C, boiling point: 56.3°C, density: 0.783) is the simplest ketone and is a colorless liquid that is miscible in all proportions with water, alcohol, or ether.

There are two major processes for the production of acetone (2-propanone). The feedstock for these is either *iso*-propyl alcohol [$(CH_3)_2CHOH$] or cumene [*iso*-propyl benzene, $C_6H_5CH(CH_3)_2$]. In the last few years there has been a steady trend away from *iso*-propyl alcohol and toward cumene, but *iso*-propyl alcohol should continue as a precursor since manufacture of acetone from only cumene would require a balancing of the market with the coproduct phenol from this process.

Acetone is made from *iso*-propyl alcohol by either dehydrogenation (preferred) or air oxidation. These are catalytic processes at 500°C and 40 to 50 psi. The acetone is purified by distillation, boiling point 56°C and the conversion per pass is 70 to 85 percent, with the overall yield being in excess of 90 percent.

$$CH_3CH(OH)CH_3 \rightarrow CH_3C(=O)CH_3 + H_2$$

$$2CH_3CH(OH)CH_3 + O_2 \rightarrow CH_3C(=O)CH_3 + 2H_2O$$

Cumene is also used as a feedstock for the production of acetone. In this process, cumene first is oxidized to cumene hydroperoxide followed by the decomposition of the cumene hydroperoxide into acetone and phenol.

The hydroperoxide is made by reaction of cumene with oxygen at 110 to 115°C until 20 to 25 percent of the hydroperoxide is formed. Concentration of the hydroperoxide to 80% is followed by catalyzed rearrangement under moderate pressure at 70 to 100°C. During the reaction, the palladium chloride ($PdCl_2$) catalyst is reduced to elemental palladium to produce hydrogen chloride that catalyzes the rearrangement, and reoxidation of the palladium is brought about by use of cupric chloride ($CuCl_2$) that is converted to cuprous chloride ($CuCl$). The cuprous chloride is reoxidized during the catalyst regeneration cycle.

The overall yield is 90 to 92 percent. By-products are acetophenone, 2-phenylpropan-2-ol, and α-methylstyrene. Acetone is distilled first at boiling point 56°C.

Vacuum distillation recovers the unreacted cumene and yields α–methylstyrene, which can be hydrogenated back to cumene and recycled. Further distillation separates phenol, boiling point 181°C, and acetophenone, boiling point 202°C.

In older industrial processes, acetone is prepared (1) by passing the vapors of acetic acid over heated lime. Calcium acetate is produced in the first step followed by a breakdown of the acetate into acetone and calcium carbonate:

$$CH_3CO_2H + CaO \rightarrow (CH_3CO_2)_2Ca + H_2O$$

$$(CH_3CO_2)_2Ca \rightarrow CH_3COCH_3 + CaCO_3$$

and (2) by fermentation of starches, such as maize, which produce acetone along with butyl alcohol.

Acetone is a very important solvent and is widely used in the manufacture of plastics and lacquers. For storage purposes, acetone may be used as a solvent for acetylene. Acetone is the starting ingredient or intermediate for numerous organic syntheses. Closely related, industrially important compounds are diacetone alcohol [$CH_3COCH_2COH(CH_3)_2$], which is used as a solvent for cellulose acetate and nitrocellulose, as well as for various resins and gums, and as a thinner for lacquers and inking materials.

Acetone is used for the production of methyl methacrylate, solvents, bisphenol A, aldol chemicals, and pharmaceuticals.

Methyl methacrylate is manufactured and then polymerized to poly(methyl methacrylate), an important plastic known for its clarity and used as a glass substitute.

Aldol chemicals refer to a variety of substances desired from acetone involving an aldol condensation in a portion of their synthesis. The most important of these chemicals is methyl *iso*-butyl ketone (MIBK), a common solvent for many plastics, pesticides, adhesives, and pharmaceuticals.

Bisphenol A is manufactured by a reaction between phenol and acetone, the two products from the cumene hydroperoxide rearrangement. Bisphenol A is an important diol monomer used in the synthesis of polycarbonates and epoxy resins.

A product known as synthetic methyl acetone is prepared by mixing acetone (50%), methyl acetate (30%), and methyl alcohol (20%) and is used widely for coagulating latex and in paint removers and lacquers.

ACETONE CYANOHYDRIN

Acetone cyanohydrin is manufactured by the direct reaction of hydrogen cyanide with acetone catalyzed by base, generally in a continuous process.

$$CH_3COCH_3 + HCN \rightarrow CH_3C(OH.CN)CH_3$$

Acetone cyanohydrin is an intermediate in the manufacture of methyl methacrylate.

ACETOPHENETIDINE

Acetophenetidine (phenacetin), an analgesic and antipyretic, is the ethyl ether of acetaminophen and is prepared from *p*-ethoxyaniline.

ACETYLENE

Acetylene (CH≡CH, ethyne or ethine, melting point –81.5°C, boiling point –84°C) is an extremely reactive hydrocarbon that is moderately soluble in water or alcohol and is markedly soluble in acetone (300 volumes of acetylene dissolve in 1 volume of acetone at 176 psi, 1216 kPa).

Acetylene burns when ignited in air with a luminous sooty flame, requiring a specially devised burner for illumination purposes. An explosive mixture is formed with air over a wide range (about 3 to 80% acetylene), but safe handling is improved when the gas is dissolved in acetone.

Acetylene is still manufactured by the action of calcium carbide, a product of the electric furnace.

$$CaC_2 + 2H_2O \rightarrow HC\equiv CH + Ca(OH)_2$$

and there are two principal methods for generating acetylene from calcium carbide.

The batch carbide-to-water, or wet, method takes place in a cylindrical water shell surmounted by a housing with hopper and feed facilities. The carbide is fed to the water at a measured rate until exhausted. The calcium hydroxide is discharged in the form of a lime slurry containing about 90% by weight water.

For large-scale industrial applications, dry generation, a continuous process featuring automatic feed, is used, in which 1 kg of water is used, per kilogram of carbide. The heat of the reaction is largely dissipated by water vaporization, leaving the by-product lime in a dry state, and part of this can be recycled to the carbide furnaces. Continuous agitation is necessary to prevent overheating, since the temperature should be kept below 150°C and the pressure lower than 204 kPa.

The newest methods of manufacturing acetylene are through the *pyrolysis, or cracking, of natural gas* or liquid hydrocarbon feeds. The processes of most interest include partial oxidation, using oxygen, thermal cracking, and an electric arc to supply both the high temper-attire and the energy. Acetylene is produced from the pyrolysis of naphtha in a two-stage cracking process in which both acetylene and ethylene are end products. Varying the naphtha feed rate can change the ratio of the two products. Acetylene also has been produced by a submerged-flame process from crude oil.

At 1327°C and higher, acetylene is more stable than other hydrocarbons but decomposes into its elements. Hence conversion, or splitting, time must be incredibly short (milliseconds). The amount of energy needed is very large and in the region of the favorable free energy.

$$2CH_4 \rightarrow HC\equiv CH + 3H_2$$

However, the decomposition of methane (CH_4) into its elements starts at 578°C, hence competes with its degradation to acetylene.

$$CH_4 \rightarrow C + 2H_2$$

To lessen this degradation after raising the methane (or other hydrocarbon) to a high temperature of about 1500°C for milliseconds, the reaction mass must be water quenched almost instantaneously.

The partial combustion (partial oxidation) of natural gas (Fig. 1) is probably the most widely used method of producing acetylene. The overall reaction of the methane (combustion and splitting) is 90 to 95 percent whereas the oxygen is 100 percent converted. The residence time is 0.001 to 0.01 seconds. The acetylene and gases are cooled rapidly by quench oil or water sprays to 38°C and have the following typical composition (percent by volume: acetylene, 8 to 10; hydrogen, 50 to 60; methane, 5; carbon monoxide, 20 to 25; and carbon dioxide, <5. The soot is removed in a carbon filter and the clean gases are compressed to 165 psi (1.14 MPa).

Acetylene is removed in a column (packed) by a selective solvent such as dimethylformamide. Carbon dioxide is flashed and stripped overhead out

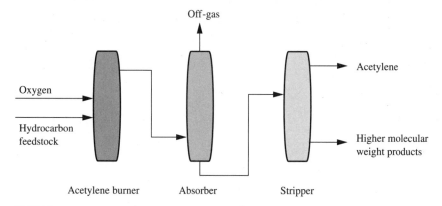

FIGURE 1   Acetylene manufacture by partial oxidation of hydrocarbons.

of the rich solvent in a column (packed), where the acetylene is fractionated out, giving a >99% by volume pure product with a 30 to 40 percent yield from the carbon in the natural gas.

Acetylene is principally used as a chemical intermediate. Acetylene reacts:

1. With chlorine, to form acetylene tetrachloride ($CHCl_2CHCl_2$) or acetylene dichloride ($CHCl=CHCl$

2. With bromine, to form acetylene tetrabromide ($CHBr_2CHBr_2$) or acetylene dibromide ($CHBr=CHBr$)

3. With hydrogen chloride (hydrogen bromide or hydrogen iodide), to form ethylene monochloride ($CH_2=CHCl$) (monobromide, monoiodide), and 1,1-dichloroethane (ethylidene chloride (CH_3CHCl_2) (dibromide, diiodide)

4. With water in the presence of a catalyst such as mercuric sulfate ($HgSO_4$) to form acetaldehyde (CH_3CHO)

5. With hydrogen, in the presence of a catalyst such as finely divided nickel (Ni) heated, to form ethylene ($CH_2=CH_2$) or ethane (CH_3CH_3)

6. With metals, such as copper (Cu) nickel (Ni), when moist, also lead (Pb) or zinc (Zn), when moist and unpurified; tin (Sn) is not attacked but sodium yields, upon heating, sodium acetylide ($CH\equiv CNa$) and disodium acetylide ($NaC\equiv CNa$)

7. With ammoniocuprous (or silver) salt solution, to form cuprous (or silver) acetylide ($HC\equiv CCu$ or $HC\equiv CAg$) which is explosive when dry and yields acetylene by treatment with acid

8. With mercuric chloride ($HgCl_2$) solution, to form trichloromercuric acetaldehyde [$C(HgCl)_3 \cdot CHO$], which yields acetaldehyde (CH_3CHO) plus mercuric chloride when treated with hydrogen chloride

ACROLEIN

Acrolein (2-propenal, CH_2=CHCHO, freezing point: $-87^{\circ}C$, boiling point: $52.7^{\circ}C$, density: 0.8427, flash point: $-18^{\circ}C$) is the simplest unsaturated aldehyde. The primary characteristic of acrolein is its high reactivity due to conjugation of the carbonyl group with a vinyl group.

Acrolein is a highly toxic material with extreme lachrymatory properties. At room temperature acrolein is a liquid with volatility and flammability somewhat similar to those of acetone, but, unlike acetone, its solubility in water is limited. Commercially, acrolein is always stored with hydroquinone and acetic acid as inhibitors.

The first commercial process for manufacturing acrolein was based on the vaporphase condensation of acetaldehyde and formaldehyde.

$$HCH=O + CH_3CH=O \rightarrow CH_2=CHCHO + H_2O$$

Catalyst developments led to a vapor-phase processes for the production of acrolein in which propylene was the starting material.

$$CH_3CH=CH_2 + [O] \rightarrow CH_2=CHCHO + H_2O$$

The catalytic vapor-phase oxidation of propylene (Fig. 1) is generally carried out in a fixed-bed multitube reactor at near atmospheric pressures and elevated temperatures (ca $350^{\circ}C$); molten salt is used for temperature control. Air is commonly used as the oxygen source and steam is added to suppress the formation of flammable gas mixtures. Operation can be single pass or a recycle stream may be employed.

The reactor effluent gases are cooled to condense and separate the acrolein from unreacted propylene, oxygen, and other low-boiling components (predominantly nitrogen). This is commonly accomplished in two absorption steps where (1) aqueous acrylic acid (CH_2=$CHCO_2H$) is condensed from the reaction effluent and absorbed in a water-based stream and (2) acrolein is condensed and absorbed in water to separate it from the propylene, nitrogen, oxygen, and carbon oxides. Acrylic acid may be recovered from the aqueous product stream if desired. Subsequent distilla-

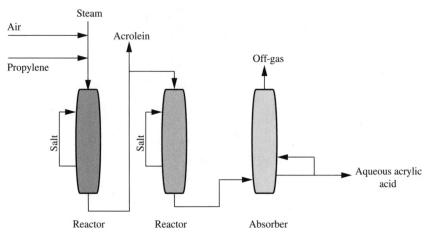

FIGURE 1   Manufacture of acrolein and acrylic acid by oxidation of propylene.

tion refining steps separate water and acetaldehyde (CH₃CHO) from the crude acrolein. In another distillation column, refined acrolein is recovered as an azeotrope with water.

The principal side reactions produce acrylic acid, acetaldehyde, acetic acid, carbon monoxide, and carbon dioxide, and a variety of other aldehydes and acids are also formed in small amounts.

ACRYLIC ACID

Acrylic acid (CH_2=$CHCO_2H$, melting point: 13.5°C, boiling point: 141°C, density: 1.045, flash point: 68°C) and acrylates were once prepared by reaction of acetylene and carbon monoxide with water or an alcohol, with nickel carbonyl as catalyst.

$$HC{\equiv}CH + CO + H_2O \rightarrow CH_2{=}CHCO_2H$$

In the presence of such catalysts as a solution of cuprous and ammonium chlorides, hydrogen cyanide adds to acetylene to give acrylonitrile (CH_2=$CHCN$). However, this process has been replaced by processes involving ammoxidation of propylene. Similarly, the process for the manufacture of acrylic acid has been superseded by processes involving oxidation of propylene (Fig. 1) although, for some countries, acetylene may still be used in acrylate manufacture.

Thus, acrylic acid is made by the oxidation of propylene to acrolein and further oxidation to acrylic acid.

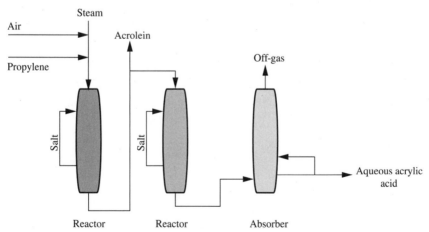

FIGURE 1 Manufacture of acrylic acid by the oxidation of propylene.

$$2CH_2{=}CHCH_3 + O_2 \rightarrow 2CH_2{=}CHCH{=}O$$

$$2CH_2{=}CHCHO + O_2 \rightarrow 2CH_2{=}CHCOOH$$

Another method of acrylic acid production is by the hydrolysis of acrylonitrile:

$$CH_2{=}CH{-}CN + 2H_2O + H^+ \rightarrow CH_2{=}CH{-}COOH + NH_4^+$$

Acrylic acid and its salts are raw materials for an important range of esters, including methyl acrylate, ethyl acrylate, butyl acrylate, and 2-ethylhexyl acrylate. The acid and its esters are used in polyacrylic acid and salts (including superabsorbent polymers, detergents, water treatment chemicals, and dispersants), surface coatings, adhesives and sealants, textiles, and plastic modifiers.

ACRYLIC RESINS

The methyl, ethyl, and butyl esters of acrylic and methacrylic acids are polymerized under the influence of heat, light, and peroxides. The polymerization reaction is exothermic and may be carried out in bulk for castings, or by emulsion, or in solution. The molecular weight decreases as the temperature and catalyst concentration are increased. The polymers are noncrystalline and thus very clear. Such resins are widely used because of their clarity, brilliance, ease of forming, and light weight. They have excellent optical properties and are used for camera, instrument, and spectacle lenses.

Because of their excellent dielectric strength they are often used for high-voltage line spacers and cable clamps. Emulsions are widely applied as textile finishes and paints.

ACRYLONITRILE

Acrylonitrile (2-propenonitrile, propene nitrile, vinyl cyanide, $CH_2=CHCN$; freezing point: $-83.5°C$, boiling point: $77.3°C$, density: 0.806) used to be manufactured completely from acetylene by reaction with hydrogen cyanide.

$$HC\equiv CH + HCN \rightarrow CH_2=CHCN$$

There was also a process using ethylene oxide as the starting material through addition of hydrogen cyanide (HCN) and elimination of water.

$$\overline{CH_2CH_2O} + HCN \rightarrow HOCH_2CH_2CN$$

$$HOCH_2CH_2CN \rightarrow CH_2=CHCN + H_2O$$

The presently used process focuses on the ammoxidation (ammonoxidation or oxyamination) of propylene that involves reaction of propylene, ammonia, and oxygen at 400 to 450°C and 7 to 29 psi (48 to 200 kPa) in a fluidized bed $Bi_2O_3 \cdot nMnO_3$ catalyst (Fig.1).

$$2CH_2=CHCH_3 + 2NH_3 + 3O_2 \rightarrow 2CH_2=CHCN + 6H_2O$$

The effluent is scrubbed in a countercurrent absorber and the acrylonitrile is purified by fractionation.

In one version of this process, the starting ingredients are mixed with steam and preheated before being fed to the reactor. There are two main by-products, acetonitrile (CH_3CN) and hydrogen cyanide (HCN), with accompanying formation of small quantities of acrolein ($CH_2=CHCHO$), acetone (CH_3COCH_3), and acetaldehyde (CH_3CHO). The acrylonitrile is separated from the other materials in a series of fractionation and absorption operations. A number of catalysts have been used, including phosphorus, molybdenum, bismuth, antimony, tin, and cobalt.

The most important uses of acrylonitrile are in the polymerization to polyacrylonitrile. This substance and its copolymers make good synthetic fibers for the textile industry. Acrylic is the fourth-largest synthetic fiber

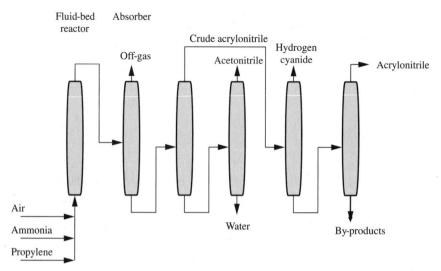

FIGURE 1 Manufacture of acrylonitrile by the ammoxidation of propylene.

produced, behind polyester, nylon, and polyolefin. It is known primarily for its warmth, similar to that natural and very expensive fiber, wool. Acrylonitrile is also used to produce plastics, including the copolymer of styrene-acrylonitrile (SA) and the terpolymer of acrylonitrile, butadiene, and styrene (ABS).

Hydrogen cyanide, a by-product of acrylonitrile manufacture, has its primary use in the manufacture of methyl methacrylate by reaction with acetone.

ADIPIC ACID

Adipic acid (melting point: 152.1°C, density: 1.344) is manufactured pre-dominantly by the oxidation of cyclohexane followed by oxidation of the cyclohexanol/cyclohexanone mixture with nitric acid (Figs. 1 and 2):

$$C_6H_{12} \rightarrow CH_2CH_2CH_2CH_2CH_2CHOH + CH_2CH_2CH_2CH_2CH_2C=O$$

$$CH_2CH_2CH_2CH_2CH_2CHOH + [O] \rightarrow HO_2C(CH_2)_4CO_2H$$

$$CH_2CH_2CH_2CH_2CH_2C=O + [O] \rightarrow HO_2C(CH_2)_4CO_2H$$

There is no need to separate the cyclohexanol/cyclohexanone mixture into its individual components; oxidation of the mixture is carried out directly.

Adipic acid can also be made by hydrogenation of phenol with a palladium or nickel catalyst (150°C, 50 psi) to the mixed oil, then nitric acid oxidation to adipic acid. If palladium is used, more cyclohexanone is formed.

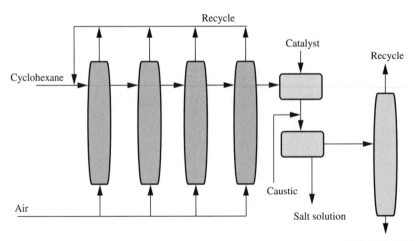

FIGURE 1 Manufacture of adipic acid by aerial oxidation of cyclohexane.

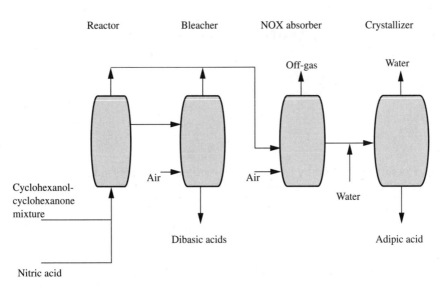

FIGURE 2   Adipic acid manufacture by nitric acid oxidation of cyclohexane

Although the phenol route for making adipic acid is not economically advantageous because phenol is more expensive than benzene, the phenol conversion to greater cyclohexanone percentages can be used successfully for caprolactam manufacture, where cyclohexanone is necessary.

Adipic acid is used to make nylon 6,6 fibers and nylon 6,6 resins; it is also used in the manufacture of polyurethanes and plasticizers.

Other starting materials for adipic acid include butadiene and 1,4-disubstituted-2-butene, which involves dicarbonylation with palladium chloride. Polar, aprotic, and nonbasic solvents are preferred for this reaction to avoid unwanted side products from hydrogenolysis or isomerization.

ADIPONITRILE

Adiponitrile is made by two different methods. One method is by the electrohydrodimerization of acrylonitrile. It is converted into hexamethylenediamine (HMDA) that is used to make nylon.

$$2CH_2{=}CHCN \rightarrow NC(CH_2)_4CN$$

$$NC(CH_2)_4CN \rightarrow H_2N(CH_2)_6NH2$$

In the electrodimerization of acrylonitrile, a two-phase system containing a phase transfer catalyst tetrabutylammonium tosylate is used.

ALCOHOLS, LINEAR, ETHOXYLATED

Ethoxylated linear alcohols can be made by the reaction of straight-chain alcohols, usually C_{12} to C_{14}, with three to seven moles of ethylene oxide.

$$C_{14}H_{29}OH + n\overline{CH_2CH_2O} \rightarrow C_{14}H_{29}O(CH_2CH_2O)_nH$$

The resulting alcohols are one type of many alcohols used for detergents. The linear alcohols can be produced from n-paraffins by way of alpha olefins or by way of the chloroparaffins. Or they can be made from alpha olefins formed from Ziegler oligomerization of ethylene.

Sulfonation and sodium salt formation of these alcohols converts them into detergents for shampoos and for dishwashing.

ALKANOLAMINES

Alkanolamines are compounds that contain both the hydroxyl (alcoholic) function (-OH) and the amino function ($-NH_2$).

Ethylene oxide, propylene oxide, or butylene oxide react with ammonia to produce alkanolamines. The more popular ethanolamines [$NH_{3-n}(C_2H_4OH)_n$, where $n = 1,2,3$: monoethanolamine, diethanolamine, and triethanolamine], are derived from the reaction of ammonia with ethylene oxide.

Alkanolamines are manufactured from the corresponding oxide and ammonia. Anhydrous or aqueous ammonia may be used, although anhydrous ammonia is typically used to favor monoalkanolamine production and requires high temperature and pressure (Fig. 1).

Isopropanolamines, $NH_{3-n}(CH_2CHOHCH_3)$, result from the reaction of ammonia with propylene oxide. Secondary butanolamines, $NH_{3-n}(CH_2CHOHCH_2CH_3)$, are the result of the reaction of ammonia with butylene oxide. Mixed alkanolamines can be produced from a mix-

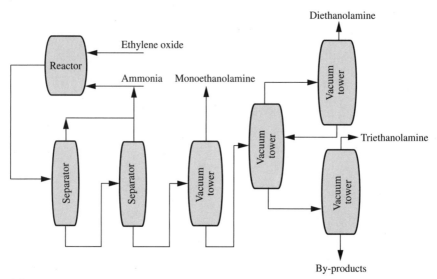

FIGURE 1 Manufacture of ethanolamines from ethylene oxide and ammonia.

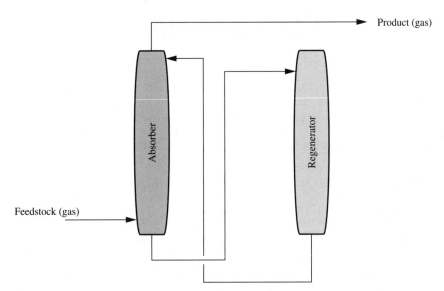

FIGURE 2   Gas cleaning using an aqueous alkanolamine.

ture of oxides reacting with ammonia. A variety of substituted alkanolamines can also be made by reaction of oxide with the appropriate
amine.

Alkanolamines are used for gas cleaning (i.e., to remove carbon dioxide
and hydrogen sulfide from gas streams) (Fig. 2), particularly in the petroleum and natural gas industries.

ALKYD RESINS

The term *alkyd resins* represents a broad class of compounds commonly used in coatings and is a particular type of polyester formed by the reaction of polyhydric alcohols and polybasic acids.

Alkyd resins are available in several forms of which the major forms are: (1) fibrous, in which the resins are compounded with long glass fibers (about ½ inch; 12 mm) and have medium strength; (2) rope, which is a medium-impact material and conveniently handled and processed; and (3) granular, in which the resins are compounded with other fibers, such as glass, asbestos, and cellulose (length about ⅟₁₆ inch; 2 mm). A commonly used member of the alkyd resin family is made from phthalic anhydride and glycerol. These resins are hard and possess very good stability. Where maleic acid is used as a starting ingredient, the resin has a higher melting point. Use of azelaic acid produces a softer and less brittle resin. Very tough and stable alkyds result from the use of adipic and other long-chain dibasic acids. Pentaerythritol may be substituted for glycerol as a starting ingredient.

The most common method of preparation of alkyd resins is the *fatty acid* method in which a glyceride oil is catalytically treated with glycerol at 225 to 250°C. The glyceride oil is simultaneously esterified and deesterified to a monoglyceride.

The esterification of a polybasic acid with a polyhydric alcohol yields a thermosetting hydroxycarboxylic resin, commonly referred to as an *alkyd resin.*

Alkyd resins are also polyesters containing unsaturation that can be cross-linked in the presence of an initiator known traditionally as a *drier.* A common example is the alkyd formed from phthalic anhydride and a glyceride of linolenic acid obtained from various plants. Cross-linking of the multiple bonds in the long unsaturated chain produces the thermoset polymer.

The processing equipment (reaction kettle and blending tank) used for unsaturated polyesters can also be used for manufacturing alkyd resins.

Alkyd resins are extensively used in paints and coatings. Some advantages include good gloss retention and fast drying characteristics. However, most unmodified alkyds have low chemical and alkali resistance. Modification with esterified rosin and phenolic resins improves hardness and chemical resistance. Styrene and vinyl toluene improve hardness and toughness. For high-temperature coatings (up to about 230°C), copolymers of silicones and alkyds are used. Such coatings include stove and heating equipment finishes. To obtain a good initial gloss, improved adhesion, and exterior durability, acrylic monomers can be copolymerized with oils to modify alkyd resins.

ALKYLBENZENES, LINEAR

Linear alkylbenzenes are made from *n*-paraffins (C_{10} to C_{14}) by either partial dehydrogenation to olefins and addition to benzene with hydrogen fluoride (HF) as catalyst or by chlorination of the paraffins and Friedel-Crafts reaction with benzene and an aluminum chloride catalyst.

In one process (Fig. 1), linear paraffins are dehydrogenated to linear olefins that are then reacted with benzene over a solid heterogeneous catalyst to produce the linear alkyl benzenes. Usually, the paraffins are of the C_{10} to C_{14} chain length.

The major uses of linear alkylbenzenes are in the manufacture of linear alkyl sulfonates that are used for manufacture of household detergents and industrial cleaners.

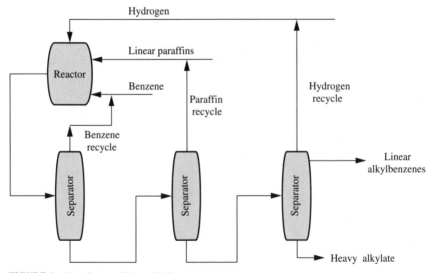

FIGURE 1   Manufacture of linear alkylbenzenes.

ALLYL ALCOHOL

Allyl alcohol (2-propen-1-ol, $CH_2=CHCH_2OH$, boiling point: 96.9°, density: 0.8520, flash point: 25°C) is the simplest unsaturated alcohol and is a colorless corrosive liquid with a pungent odor. The vapor can cause severe irritation and injury to eyes, nose, throat, and lungs. Allyl alcohol is miscible with water and miscible with many polar organic solvents and aromatic hydrocarbons, but is not miscible with n-hexane. It forms an azeotropic mixture with water and a ternary azeotropic mixture with water and organic solvents.

There are four processes for industrial production of allyl alcohol. One involves the alkaline hydrolysis of allyl chloride.

$$CH_2=CHCH_2Cl \ (+ \ OH^-) \ \rightarrow \ CH_2=CH–CH_2OH \ (+ \ Cl^-)$$

In this process, the amount of allyl chloride, 20 wt % aqueous sodium hydroxide (NaOH) solution, water, and steam are controlled as they are added to the reactor and the hydrolysis is carried out at 150°C, 200 psi (1.4 MPa) and pH 10 to 12. Under these conditions, conversion of allyl chloride is near quantitative (97 to 98 percent), and allyl alcohol is selectively produced in 92 to 93 percent yield. The main by-product is diallyl ether ($CH_2=CHCH_2OCH_2CH=CH_2$). At high alkali concentrations, the amount of by-product, diallyl ether, increases, and at low concentrations, conversion of allyl chloride does not increase.

A second process has two steps. The first step is oxidation of propylene to acrolein and the second step is reduction of acrolein to allyl alcohol by a hydrogen transfer reaction, using isopropyl alcohol.

$$CH_3CH=CH_2 + O_2 \ \rightarrow \ CH_2=CHCH=O + H_2O$$

$$CH_2=CHCHO + (CH_3)_2CHOH \ \rightarrow \ CH_2=CHCH_2OH + CH_3COCH_3$$

Another process is isomerization of propylene oxide in the presence of a catalyst (lithium phosphate, Li_3PO_4).

$$CH_3 \overline{CHCH_2O} \ \rightarrow \ CH_2=CHCH_2OH$$

   In this process, the fine powder of lithium phosphate used as catalyst is dispersed, and propylene oxide is fed at 300°C to the reactor, and the product, allyl alcohol, together with unreacted propylene oxide is removed by distillation. By-products such as acetone and propionaldehyde, which are isomers of propylene oxide, are formed, but the conversion of propylene oxide is 40 percent and the selectivity to allyl alcohol reaches more than 90 percent. Allyl alcohol obtained by this process may contain small amounts (<1%) of propanol.

   The fourth process for the production of allyl alcohol was developed partly for the purpose of producing epichlorohydrin via allyl alcohol as the intermediate, using a palladium catalyst.

$$CH_3CH=CH_2 + CH_3COOH + O_2 \rightarrow CH_2=CHOCOCH_3$$

$$CH_2=CHOCOCH_3 + H_2O \rightarrow CH_2=CHCH_2OH + CH_3CO_2H$$

   In the first step of the process (Fig. 1), the acetoxylation of propylene is carried out in the gas phase, using solid catalyst containing palladium as the main catalyst at 160 to 180°C and 70 to 140 psi (0.49 to 0.98 MPa). The reactor effluents from the reactor are separated into liquid components and gas components. The liquid components containing allyl acetate are sent to the hydrolysis process. The gas components contain unreacted gases and

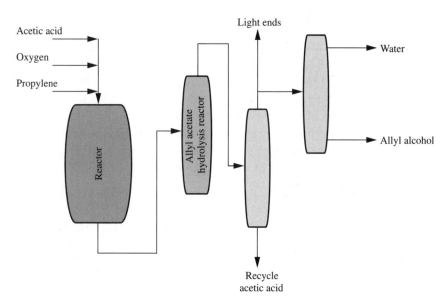

FIGURE 1   Manufacture of allyl alcohol (via allyl acetate) from propylene and acetic acid.

carbon dioxide and, after removal of the carbon dioxide, the unreacted gases are recycled to the reactor.

In the second step, the hydrolysis, which is an equilibrium reaction of allyl acetate, an acid catalyst is used and the reaction takes place at 60 to 80°C and allyl alcohol is selectively produced in almost 100 percent yield. Acetic acid recovered from the hydrolysis process, is reused in the first step.

Allyl alcohol forms an azeotropic mixture with water, and the mixture is a homogeneous liquid. Therefore, to obtain dry allyl alcohol, ternary azeotropic distillation and dehydration are required.

Allyl alcohol exhibits both bacterial and fungicidal effects and has been used as such or as a source for derivatives with these effects.

ALUMINA

Pure alumina (Al_2O_3) is a dry, snow-white, free-flowing crystalline powder and may be obtained in a wide range of particle sizes.

There are two main types of alumina (bauxite) ores used as the primary sources for aluminum metal and aluminum chemicals: aluminum hydroxide [$Al(OH)_3$] (gibbsite) and a mixed aluminum oxide hydroxide [$AlO(OH)$] (boehmite). Thus, *bauxite* is a term for a family of ores rather than a substance of one definite composition. An average composition of the ores used by industry today would be: alumina (Al_2O_3), 35 to 60%; silica (SiO_2), 1 to 15%; ferric oxide (Fe_2O_3), 5 to 40%; and titanium dioxide (TiO_2), 1 to 4%.

In the process to produce alumina (Fig. 1), bauxite is crushed and wet ground to 100-mesh, dissolved under pressure and heated in digesters with concentrated spent caustic soda solution from a previous cycle and sufficient lime and soda ash. Sodium aluminate is formed, and the dissolved silica is precipitated as sodium aluminum silicate. The undissolved residue (red mud) is separated from the alumina solution by filtration and washing and sent to recovery. Thickeners and Kelly or drum filters are used. The filtered solution of sodium aluminate is hydrolyzed to precipitate aluminum hydroxide by cooling. The precipitate is filtered from the liquor, washed, and heated to 980°C in a rotary kiln to calcine the aluminum hydroxide.

Several other processes for producing alumina based on ores other than bauxite have been announced. One process uses alunite, a hydrous sulfate of aluminum and potassium. It is claimed to be capable of producing 99% pure alumina from alunite containing only 10 to 15% alumina, compared with bauxite that assays 50% alumina. The alunite is crushed, dehydroxylated by heating to 750°C, ground, and treated with aqueous ammonia. Filtration removes the alumina hydrate, and potassium and aluminum sulfates are recovered from the filtrate (to be used as fertilizer constituents). The alumina hydrate is treated with sulfur dioxide gas, and the resulting aluminum sulfate is converted to alumina by heating in a kiln.

Another process for alumina manufacture involves treatment of clay and shale with concentrated sulfuric acid. Hydrochloric acid is added dur-

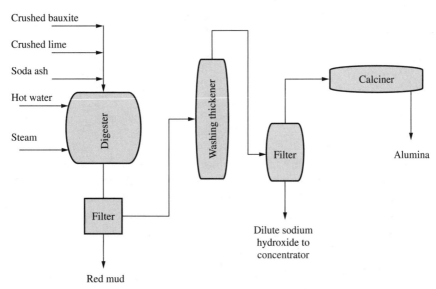

FIGURE 1 Manufacture of alumina.

ing the crystallization step to form aluminum chloride which crystallizes readily.

Other processes involve the treatment of clay with nitric acid and the continuous electrolysis of aluminum chloride.

Alumina is used to produce aluminum by the electrolytic process, and the purity of the aluminum is determined mainly by the purity of the alumina used. Thus, commercial grades of alumina are 99 to 99.5% pure with traces of water, silica (SiO_2), ferric oxide (Fe_2O_3), titanium dioxide (TiO_2), and (ZnO), and very minute quantities of other metal oxides.

Other uses include manufacture of a variety of aluminum salts water purification, glassmaking, production of steel alloys, waterproofing of textiles, coatings for ceramics, abrasives and refractory materials, cosmetics, electronics, drying gases and dehydrating liquids (such as alcohol, benzol, carbon tetrachloride, ethyl acetate, gasoline, toluol, and vegetable and animal oils), filter aids in the manufacture of lubricating and other oil products, catalysts for numerous reactions, polishing compounds, and linings for high-temperature furnaces.

ALUMINUM

Aluminum (melting point: 660°C, boiling point: 2494°C) is the most abundant metal in the world and makes up 7 to 10% by weight of the earth's crust.

Aluminum is manufactured by the electrolytic reduction of pure alumina (Al_2O_3) in a bath of fused cryolite (Na_3AlF_6). It is not possible to reduce alumina with carbon because aluminum carbide (Al_4C_3) is formed and a back-reaction between aluminum vapor and carbon dioxide in the condenser quickly reforms the original aluminum oxide again.

The electrolytic cells are large containers (usually steel), and each is a cathode compartment lined with either a mixture of pitch and anthracite coal or coke baked in place by the passage of electric current or prebaked cathode blocks cemented together.

Two types of cells are used in the Hall-Heroult process: those with multiple prebaked anodes (Fig. 1), and those with a self-baking, or Soderberg, anode. In both types of cell, the anodes are suspended from above and are connected to a movable anode bus so that their vertical position can be adjusted. The prebaked anode blocks are manufactured from a mixture of low-ash calcined petroleum coke and pitch or tar formed in hydraulic presses, and baked at up to 1100°C.

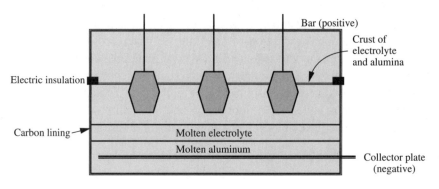

FIGURE 1   Manufacture of aluminum.

ALUMINUM CHLORIDE

Aluminum chloride ($AlCl_3$) is a white solid when pure that sublimes on heating and, in the presence of moisture, decomposes with the evolution of hydrogen chloride.

Anhydrous aluminum chloride is manufactured primarily by the reaction of chlorine vapor on molten aluminum.

$$2Al + 3Cl_2 \rightarrow 2AlCl_3$$

In the process, chlorine is fed in below the surface of the aluminum, and the product sublimes and is collected by condensing. These air-cooled condensers are thin-walled, vertical steel cylinders with conical bottoms. Aluminum chloride crystals form on the condenser walls and are periodically removed, crushed, screened, and packaged in steel containers.

Aluminum chloride is used in the petroleum industries and various aspects of organic chemistry technology. For example, aluminum chloride is a catalyst in the alkylation of paraffins and aromatic hydrocarbons by olefins and also in the formation of complex ketones, aldehydes, and carboxylic acid derivatives.

ALUMINUM SULFATE

Aluminum sulfate [$Al_2(SO_4)_3$, alum, filter alum, papermaker's alum] is manufactured from aluminum oxide (Al_2O_3, alumina, bauxite). A mixture of the crude ore and sulfuric acid is heated at 105 to 110°C for 15 to 20 hours.

$$Al_2O_3 \cdot 2H_2O + 3H_2SO_4 \rightarrow Al_2(SO_4)_3 + 5H_2O$$

Filtration of the aqueous solution is followed by evaporation of the water to give the product, which is processed into a white powder.

Alum has two prime uses. It is bought by the pulp and paper industry for coagulating and coating pulp fibers into a hard paper surface by reacting with small amounts of sodium carboxylates (soap) present. Aluminum salts of carboxylic acids are very gelatinous. In water purification it serves as a coagulant, pH conditioner, and phosphate and bacteria remover. It reacts with alkali to give an aluminum hydroxide floc that drags down such impurities in the water. For this reason it also helps the taste of water.

$$6RCO_2^-Na^+ + Al_2(SO_4)_3 \rightarrow 2(RCO_2^-)_3Al^{3+} + 3Na_2SO_4$$

$$Al_2(SO_4)_3 + 6NaOH \rightarrow 2Al(OH)_3 + 3Na_2SO_4$$

Pharmaceutically, aluminum sulfate is employed in dilute solution as a mild astringent and antiseptic for the skin. The most important single application of aluminum sulfate is in clarifying water; sodium aluminate, which is basic, is sometimes used with aluminum sulfate, which is acid, to produce the aluminum hydroxide.

Aluminum sulfate is also used in sizing of paper, as a mordant in the dye industry, chemical manufacturing, concrete modification, soaps, greases, fire extinguishing solutions, tanning, cellulosic insulation, and in some baking powders.

AMITRIPTYLINE

Amitriptyline hydrochloride and imipramine hydrochloride are similar in structure, with the exception of the nitrogen in the center ring, and belong to the family of phenothiazine compounds. Finally, the two-carbon bridge linking the aromatic rings may be ethyl ($-CH_2CH_2-$) or ethylene ($-CH=CH-$).

These compounds are central nervous system stimulants or antidepressants although such activity is usually restricted to compounds having a two- or three-carbon side chain and methyl-substituted or unsubstituted amino groups in the side chain although derivatives with substituents on the aromatic ring may have pharmacological activity.

The synthesis of amitriptyline starts from the key intermediate dibenzosuberone (manufactured from phthalic anhydride) and can proceed by two pathways (Fig. 1). Treatment of dibenzosuberone with cyclopropyl Grignard gives the tertiary alcohol after hydrolysis. Reaction of the alco-

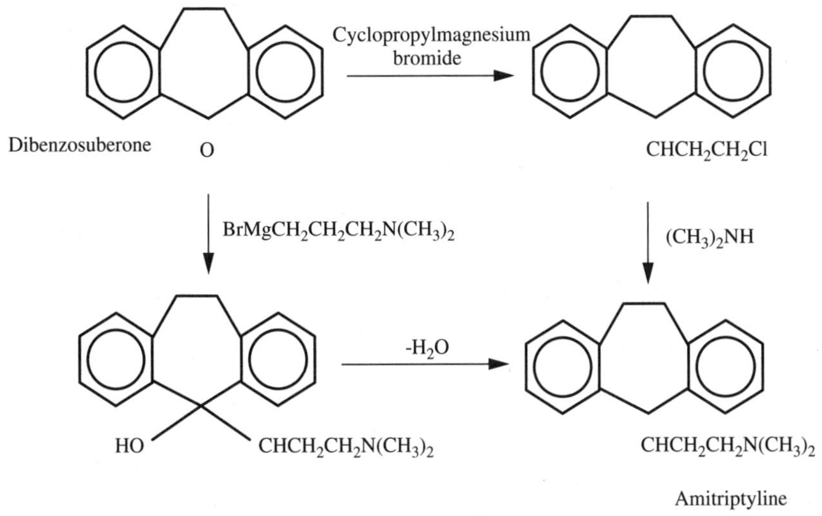

FIGURE 1   Manufacture of amitriptyline.

hol with hydrochloric acid proceeds with rearrangement and opening of the strained cyclopropane to give a chloride. Reaction and displacement of the chlorine atom with dimethylamine forms amitriptyline. Alternatively, dibenzosuberone can be reacted with dimethylaminopropyl Grignard to form an alcohol, which upon dehydration forms amitriptyline.

Amitriptyline is recommended for the treatment of mental depression, with improvement in mood seen in 2 to 3 weeks after the start of medication.

AMMONIA

Ammonia (NH_3, melting point $-77.7°C$, boiling point $-33.4°C$, and density 0.817 at $-79°C$ and 0.617 at $15°C$) is a colorless gas with a penetrating, pungent-sharp odor in small concentrations that, in heavy concentrations, produces a smothering sensation when inhaled. Ammonia is soluble in water and a saturated solution contains approximately 45% ammonia by weight at the freezing temperature of the solution and about 30% ammonia by weight at standard conditions. Ammonia dissolved in water forms a strongly alkaline solution of ammonium hydroxide (NH_4OH) and the aqueous solution is called *ammonia water, aqua ammonia*, or sometimes *ammonia* (although this is misleading). Ammonia burns with a greenish-yellow flame.

The first breakthrough in the large-scale synthesis of ammonia resulted from the development of the Haber process in 1913 in which ammonia was produced by the direct combination of two elements, nitrogen and hydrogen, in the presence of a catalyst (iron oxide with small quantities of cerium and chromium) at a relatively high temperature (550°C) and under a pressure of about 2940 psi (20.3 MPa).

$$N_2 + 3H_2 \rightarrow 2NH_3$$

In the Haber process (Fig. 1), the reaction of nitrogen and hydrogen gases is accomplished by feeding the gases to the reactor at 400 to 600°C. The reactor contains an iron oxide catalyst that reduces to a porous iron metal in the nitrogen/hydrogen mixture. Exit gases are cooled to -0 to $-20°C$, and part of the ammonia liquefies; the remaining gases are recycled.

The process for ammonia manufacture will vary somewhat with the source of hydrogen, but the majority of ammonia plants generate the hydrogen by steam reforming natural gas or hydrocarbons such as naphtha (Fig. 2).

If the hydrogen is made by steam reforming, air is introduced at the secondary reformer stage to provide nitrogen for the ammonia reaction. The

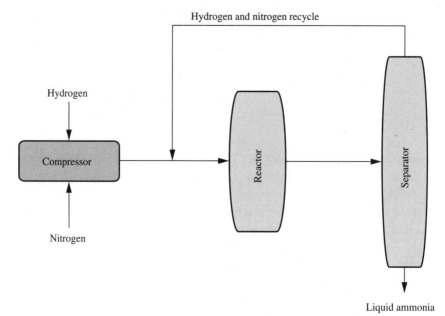

FIGURE 1 Ammonia manufacture from hydrogen and nitrogen by the Haber process.

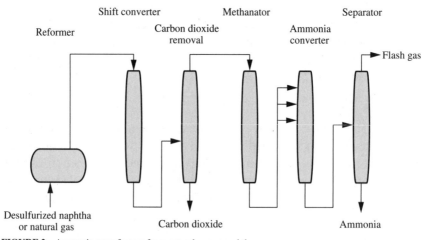

FIGURE 2 Ammonia manufacture from natural gas or naphtha.

oxygen of the air reacts with the hydrocarbon feedstock in combustion and helps to elevate the temperature of the reformer. Otherwise nitrogen can be added from liquefaction of air. In either case a hydrogen-nitrogen mixture is furnished for ammonia manufacture.

Ammonia is used for the manufacture of fertilizers or for the manufacture of other nitrogen-containing compounds used for fertilizer or, to a lesser extent, explosives, plastics, and fibers. Explosives made from ammonia are ammonium nitrate and (via nitric acid) the nitroglycerin used in dynamite. Plastics include (via urea) urea-formaldehyde and melamine-formaldehyde resins. Some ammonia ends up in fibers, since it is used to make hexamethylenediamine (HMDA), adipic acid, or caprolactam, all nylon precursors.

AMMONIUM CHLORIDE

Ammonium chloride (NH_4Cl, density 1.52) is a white crystalline solid that decomposes at 350°C and sublimes at 520°C under controlled conditions. It is also known as *sal ammoniac* and is soluble in water and in aqueous solutions of ammonia; it is slightly soluble in methyl alcohol.

Ammonium chloride is produced by neutralizing hydrochloric acid (HCl) with ammonia gas or with liquid ammonium hydroxide, and evaporating the excess water, followed by drying, crystallizing, and screening operations.

$$NH_3 + HCl \rightarrow NH_4Cl$$

Ammonium chloride can also be produced in the gaseous phase by reacting hydrogen chloride gas with ammonia.

Ammonium chloride is used as an ingredient of dry cell batteries, as a soldering flux, as a processing ingredient in textile printing and hide tanning, and as a starting material for the manufacture of other ammonium chemicals.

AMMONIUM NITRATE

Ammonium nitrate (NH_4NO_3), a colorless crystalline solid, occurs in two forms: (1) α-ammonium nitrate (tetragonal crystals, stable between –16°C and 32°C; melting point: 169.9°C; density: 1.66) and (2) β-ammonium nitrate (rhombic or monoclinic crystals, stable between 32°C and 84°C with decomposition occurring above 210°C; density: 1.725).

When heated, ammonium nitrate yields nitrous oxide (N_2O) gas and can be used as an industrial source of that gas. Ammonium nitrate is soluble in water, slightly soluble in ethyl alcohol, moderately soluble in methyl alcohol, and soluble in acetic acid solutions containing ammonia.

Ammonium nitrate is manufactured from ammonia and nitric acid.

$$NH_3 + HNO_3 \rightarrow NH_4NO_3$$

In the process (Figs. 1 and 2), the gases are fed to the reactor in which the heat of neutralization boils the mixture, concentrating it to 85% nitrate. Vacuum evaporation at 125 to 140°C further concentrates the solution to 95%. The last water of this hygroscopic material is very difficult to remove. The hot solution is pumped to the top of a spray tower 60 to 70 m high, where it is discharged through a spray head and solidifies as it falls in the air to form small spherical pellets, prills, of 2 mm diameter that are screened, further dried, and dusted with clay to minimize sticking.

If properly proportioned and preheated, the reaction can be run continuously to produce molten ammonium nitrate containing very little water (1 to 5%), which can be formed into small spheres (prills) by dropping the reaction product through a shot tower or into flakes by cooling it on belts or drums. By fluidized bed treatment, it is possible to obtain a dry granular material as product; batch processes have also been used.

Ammonium nitrate finds major applications in explosives and fertilizers, and additional uses in pyrotechnics, freezing mixtures (for obtaining low temperatures), as a slow-burning propellant for missiles (when formulated with other materials, including burning-rate catalysts), as an ingredient in rust inhibitors (especially for vapor-phase corrosion), and as a component of insecticides.

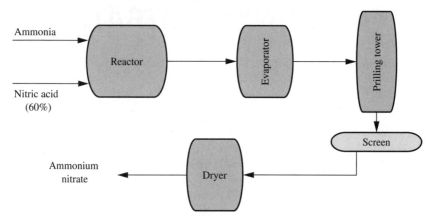

FIGURE 1   Manufacture of ammonium nitrate.

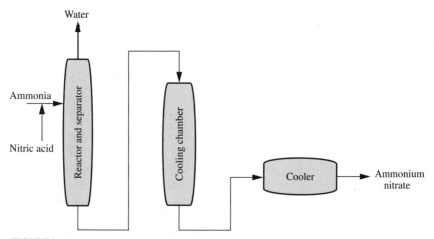

FIGURE 2   Ammonium nitrate by the Stengel process.

Ammonium nitrate is used in explosives, and many commercial and military explosives contain ammonium nitrate as the major explosive ingredient. Ammonium nitrate is difficult to detonate, but, when sensitized with oil or mixed with other explosive materials, it can be detonated with a large booster-primer. Amatol is a mixture of trinitrotoluene (TNT) and granular ammonium nitrate and is a major conventional military explosive. The explosive ANFO is a mixture of ammonium nitrate and fuel oil.

Explosive decomposition causes ammonium nitrate to rapidly and violently decompose to form elemental nitrogen.

$$NH_4NO_3 \rightarrow 2N_2 + 4H_2O + O_2$$

However, under different conditions, i.e., at 200 to 260°C, ammonium nitrate is safely decomposed to form the anesthetic nitrous oxide, and this reaction is used for the commercial manufacture of nitrous oxide.

$$NH_4NO_3 \rightarrow N_2O + 2H_2O$$

Slurry explosives consist of oxidizers (NH_4NO_3 and $NaNO_3$), fuels (coals, oils, aluminum, other carbonaceous materials), sensitizers (trinitrotoluene, nitrostarch, and smokeless powder), and water mixed with a gelling agent to form a thick, viscous explosive with excellent water-resistant properties. Slurry explosives may be manufactured as cartridged units, or mixed on site.

As a fertilizer, ammonium nitrate contains 35% nitrogen and, because of the explosive nature of the compound, precautions in handling are required. This danger can be minimized by introducing calcium carbonate into the mixture, reducing the effective nitrogen content of the product to 26% by weight.

Ammonium nitrate is hygroscopic; clay coatings and moisture-proof bags are used to preclude spoilage in storage and transportation.

AMMONIUM PHOSPHATE

There are three possible ammonium orthophosphates, only two of which are manufactured on any scale.

Monoammonium phosphate ($NH_4H_2PO_4$, white crystals, density: 1.803) is readily made by reacting ammonia with phosphoric acid, centrifuging, and drying in a rotary dryer. It is used in quick-dissolving fertilizers and as a fire-retarding agent for wood, paper, and cloth. Diammonium phosphate [$(NH_4)_2HPO_4$, white crystals, density: 1.619] requires a two-stage reactor system in order to prevent loss of ammonia. A granulation process follows with completion of the reaction in a rotary drum. Both compounds are soluble in water and insoluble in alcohol or ether.

A third compound, triammonium phosphate [$(NH_4)_3PO_4$] does not exist under normal conditions because, upon formation, it immediately decomposes, losing ammonia (NH_3) and reverts to one of the less alkaline forms.

Ammonium phosphates usually are manufactured by neutralizing phosphoric acid with ammonia in which control of the pH (acidity/alkalinity) determines which of the ammonium phosphates will be produced.

Pure grades of a particular ammonium phosphate can be produced by crystallization of solutions obtained from furnace-grade phosphoric acid.

There has been a trend toward the production of ammonium phosphates in powder form by a process in which concentrated phosphoric acid is neutralized under pressure, and the heat of neutralization is used to remove the water in a spray tower. The powdered product then is collected at the bottom of the tower. Ammonium nitrate/ammonium phosphate combination products can be obtained either by neutralizing mixed nitric acid and phosphoric acid or by the addition of ammonium phosphate to an ammonium nitrate melt.

Large quantities of the ammonium phosphates are used as fertilizers and in fertilizer formulations as well as fire retardants in wood building materials, paper and fabric products, and in matches to prevent afterglow. Solutions of the ammonium phosphates sometimes are air-dropped to retard forest fires, serving the double purpose of fire fighting and fertilizing the soil to accelerate new plant growth.

Ammonium phosphates are also used in baking powder formulations, as nutrients in the production of yeast, as nutritional supplements in animal feeds, for controlling the acidity of dye baths, and as a source of phosphorus in certain kinds of ceramics.

AMMONIUM PICRATE

Ammonium picrate (explosive D) is manufactured by the neutralization of a hot aqueous solution of picric acid (2,4,6-trinitrophenol) with aqueous ammonia.

It is used in armor-piercing shells as a bursting charge. Ammonium picrate is a salt and does not melt, so it must be loaded by compression.

AMMONIUM SULFATE

Ammonium sulfate [$(NH_4)_2SO_4$, density: 1.769] is a colorless crystalline solid that decomposes above 513°C and is soluble in water but insoluble in alcohol.

Ammonium sulfate was originally manufactured by using sulfuric acid to scrub by-product ammonia from coke-oven gas, and much is still produced in this manner. Most of the ammonium sulfate produced is now made by reaction between synthetic ammonia and sulfuric acid.

$$2NH_3 + H_2SO_4 \rightarrow (NH_4)_2SO_4$$

Where sulfur for sulfuric acid is at a premium, a process based on gypsum and carbon dioxide from combustion can be used:

$$(NH_4)_2CO_3 + CaSO_4 2H_2O \rightarrow CaCO_3 + 2H_2O + (NH_4)_2SO_4$$

Water is removed by evaporation, and the product is crystallized to large, white uniform crystals, melting point 513°C, with decomposition. Anhydrite ($CaSO_4$) can also be used in this process.

Large quantities of ammonium sulfate are produced by a variety of industrial neutralization operations required for alleviation of stream pollution by free sulfuric acid (H_2SO_4) as well as in the manufacture of caprolactam. Ammonium sulfate also is a byproduct of coke oven operations where the excess ammonia formed is neutralized with sulfuric acid to form the ammonium sulfate.

Ammonium sulfate is used predominantly as a fertilizer. As a fertilizer, $(NH_4)_2SO_4$ has the advantage of adding sulfur to the soil as well as nitrogen, as it contains 21% by weight nitrogen and 24% by weight sulfur. Ammonium sulfate also is used in electric dry cell batteries, as a soldering liquid, as a fire retardant for fabrics and other products, and as a source of certain ammonium chemicals. Other uses include water treatment, fermentation processes, fireproofing agents, and leather tanning.

ANILINE

Aniline is an important derivative of benzene that can be made in two steps by nitration to nitrobenzene and either catalytic hydrogenation or acidic metal reduction to aniline.

$$C_6H_6 + HNO_3 \rightarrow C_6H_5NO_2 + H_2O$$

$$C_6H_5NO_2 + 3H_2 \rightarrow C_6H_5NH_2 + 2H_2O$$

Both steps occur in excellent yield.

The reaction of ammonia and phenol is also being used for aniline production.

$$C_6H_5OH + NH_3 \rightarrow C_6H_5NH_2 + H_2O$$

Major uses of aniline include the manufacture of p,p'-methylene diphenyl diisocyanate (MDI), which is polymerized with a diol (HO-R-OH) to give a polyurethane. Two moles of aniline react with formaldehyde to give p,p'-methylenedianiline (MDA), which reacts with phosgene to give p,p'-methylene diphenyl diisocyanate. Toluene diisocyanate (TDI) also reacts with a diol to give a polyurethane, but polyurethanes derived from p,p'-methylene diphenyl diisocyanate are more rigid than those from toluene diisocyanate.

Aniline is used in the rubber industry for the manufacture of various vulcanization accelerators and age resistors. Aniline is also used in the production of herbicides, dyes and pigments, and specialty fibers.

See Amination.

ANISALDEHYDE

Anisaldehyde is a colorless oily liquid with an agreeable odor resembling that of coumarin, which is developed only after dilution and in mixtures.

It is made by the oxidation of anethole (the chief constituent of anise, star anise, and fennel oils).

$$CH_3OC_6H_4CH{=}CHCH_3 \rightarrow CH_3OC_6H_4CH{=}O$$

Anethole is obtained from the higher-boiling fractions of pine oil.

ANTIBIOTICS

The term *antibiotic* is broad and is defined as a substance produced by microorganisms that has the capacity of inhibiting the growth of and even of destroying other microorganisms by the action of very small amounts of the substance.

Penicillin, erythromycin, tetracycline, and cephalosporins are among the most widely used. Synthetic modifications of the naturally occurring antibiotic compounds have produced many variations that have the necessary clinical properties.

Many antibiotics are now manufactured (Fig. 1) and caution is often required during production because of the instability of these compounds to heat, variation in the pH, and chemical action, and they may even decompose in solution.

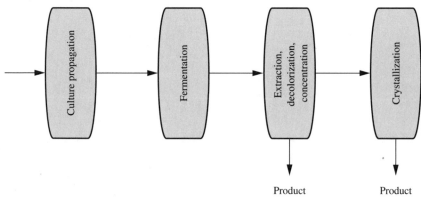

FIGURE 1   Schematic for the manufacture of antibiotics.

ANTIHISTAMINES

Antihistamines are sold under such trade names as Dimetapp®, Actifed®, and Benadryl®. They have a structure including the group R-X-C-C-N=, where X can be nitrogen, oxygen, or carbon (Fig. 1).

Antihistamines are produced by certain key steps, such as the synthesis of diphenhydramine from diphenylmethane (Fig. 2).

Antihistamines are used to alleviate allergic conditions such as rashes and runny eyes and nose and are decongestants that are used for swelled sinuses and nasal passages during the common cold. These symptoms are caused by histamine and hence the drugs that get rid of them are antihistamines. Antihistamines are also sleep inducers.

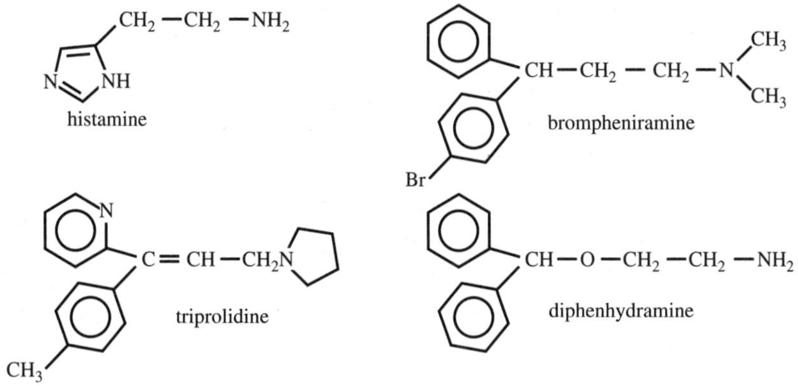

FIGURE 1   Examples of antihistamines.

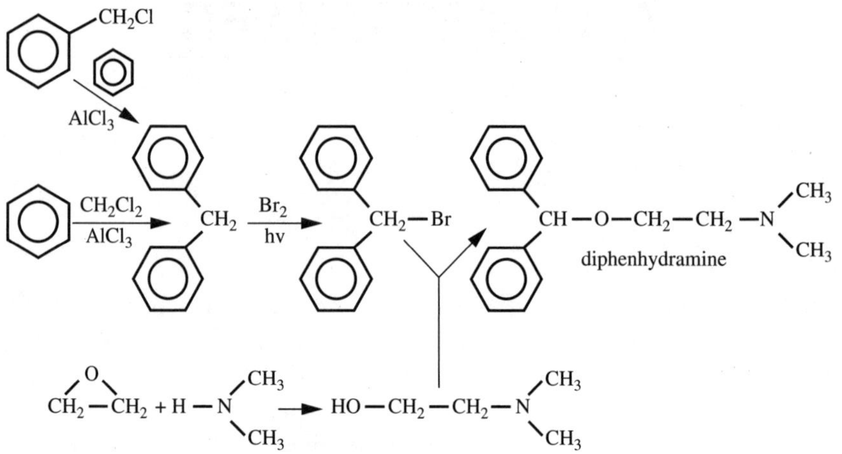

FIGURE 2   Manufacture of diphenylhydramine.

ARGON

*See **Rare Gases**.*

ASPIRIN

Aspirin (acetylsalicylic acid) is by far the most common type of analgesic, an important class of compounds that relieve pain, and it also lowers abnormally high body temperatures. Aspirin also finds use in reducing inflammation caused by rheumatic fever and rheumatoid arthritis.

The manufacture of aspirin is based on the synthesis of salicylic acid from phenol. Reaction of carbon dioxide with sodium phenoxide is an electrophilic aromatic substitution on the ortho, para-directing phenoxy ring. The ortho isomer is steam distilled away from the para isomer.

$$C_6H_5OH + CO_2 \rightarrow HOC_6H_4CO_2H$$

Salicylic acid reacts easily with acetic anhydride to give aspirin.

$$HOC_6H_4CO_2H + (CH_3CO)_2O \rightarrow CH_3OCOC_6H_4CO_2H + CH_3CO_2H$$

In this process, a 500-gallon glass-lined reactor is needed to heat the salicylic acid and acetic anhydride for 2 to 3 hours. The mixture is transferred to a crystallizing kettle and cooled to 3°C. Centrifuging and drying of the crystals yields the bulk aspirin. The excess solution is stored and the acetic acid is recovered to make more acetic anhydride.

The irritation of the stomach lining caused by aspirin can be alleviated with the use of mild bases such as sodium bicarbonate, aluminum glycinate, sodium citrate, aluminum hydroxide, or magnesium trisilicate (a trademark for this type of aspirin is Bufferin®).

Both phenacetin and the newer replacement acetaminophen are derivatives of p-aminophenol. Although these latter two are analgesics and antipyretics, the aniline-phenol derivatives show little if any antiinflammatory activity. p-Aminophenol itself is toxic, but acylation of the amino group makes it a convenient drug.

A trademark for acetaminophen is Tylenol®. Excedrin® is acetaminophen, aspirin, and caffeine. Acetaminophen is easily synthesized from phenol.

See Salicylic Acid.

BARBITAL

Barbital®, diethylbarbituric acid, is sold under the trade name Veronal®. It is the oldest of the long-acting barbiturates and is derived through diethyl malonate.

BARBITURATES

Barbiturates are sedative drugs that are derived from barbituric acid and have the ability to depress the central nervous system and act especially on the sleep center in the brain, thus their sedative and sometimes hypnotic effects.

The method of synthesis for thousands of barbital analogs involves the reaction of urea with various derivatives of malonic acid, usually a diethyl ester of a dialkyl-substituted malonic acid.

The barbiturates are usually administered as the sodium salts. The N-H bonds are acidic and although barbituric acid itself is inactive, a range of activities is obtained that varies with the groups at R and R^1. Activity and toxicity both increase with the size of the groups. Branching and unsaturation decrease the duration of action.

Barbiturates have been used in sleeping pills, but in recent years several other compounds also have been introduced for this purpose. Barbiturates induce a feeling of relaxation, usually followed by sleep, and have been used to provide temporary respite in times of unusual emotional stress, but they will only prolong the stress and patient reliance on the drugs is common.

The maximum therapeutic index (tolerated dose/minimum effective dose) is highest when the two groups have a total of 6 to 10 carbons. Major drawbacks of their use are their habit formation and their high toxicity when alcohol is present in the bloodstream.

BARIUM CARBONATE

*See **Barium Salts**.*

BARIUM SALTS

The most common naturally occurring barium compounds are the mineral carbonate (witherite, $BaCO_3$) and the sulfate (barite, $BaSO_4$).

The manufacture of soluble barium salts involves treatment of the mineral (usually barite) with the relevant acid, filtration to remove insoluble impurities, and crystallization of the salt.

The high-temperature reduction of barium sulfate with coke yields the water-soluble barium sulfide (BaS), which is subsequently leached out. Treatment of barium sulfide with the relevant chemical yields the desired barium salt. Purification of the product is complicated by the impurities introduced in the coke. Pure barium carbonate and barium sulfate are made by precipitation from solutions of water-soluble barium salts.

Barite is ground, acid-washed, and dried to produce a cheap pigment or paper or rubber filler, or changed to *blanc fixe*.

The applications of barium compounds are varied and include use as oil-drilling mud. Barium carbonate is sometimes employed as a neutralizing agent for sulfuric acid and, because both barium carbonate and barium sulfate are insoluble, no contaminating barium ions are introduced. The foregoing application is found in the synthetic dyestuff industry.

Barium carbonate is used in the glass industry and in the brick and clay industry. When barium carbonate is added to the clay used in making bricks, it immobilizes the calcium sulfate and prevents it from migrating to the surface of the bricks and producing a whitish surface discoloration.

Barium sulfate is a useful white pigment, particularly in the precipitated form, *blanc fixe*. It is used as a filler for paper, rubber, linoleum, and oilcloth. Because of its opacity to x-rays, barium sulfate, in a purified form, is important in contour photographs of the digestive tract.

The paint industry is the largest single consumer of barium compounds. Barium sulfide and zinc sulfate solutions are mixed to give a precipitate of barium sulfate and zinc sulfide, which is heat treated to yield the pigment *lithopone*. Barium chlorate and nitrate are used in pyrotechnics to impart a green flame. Barium chloride is applied where a soluble barium compound is needed.

BARIUM SULFATE

Barium sulfate ($BaSO_4$, melting point 1580°C with decomposition) occurs as colorless rhombic crystals. It is soluble in concentrated sulfuric acid, forming an acid sulfate; dilution with water reprecipitates barium sulfate. Precipitated barium sulfate, known as *blanc fixe*, is prepared from the reaction of aqueous solutions of barium sulfide (BaS) and sodium sulfate (Na_2SO_4).

Barium sulfate, often prepared by grinding the ore barite, is not usually used alone as a white pigment because of its poor covering power, but is widely used as a pigment extender. It also contributes to gloss. Its principal use (over 90 percent) is in oil-drilling muds. *Blanc fixe* is made by precipitation of a soluble barium compound, such as barium sulfide or barium chloride, by a sulfate. This form has finer particles than the ground barite and is often used in printing inks to impart transparency.

*See also **Barium Salts, Lithopone.***

BARIUM SULFIDE

*See **Barium Salts, Lithopone.***

BAUXITE

See Alumina.

BENZALDEHYDE

Benzaldehyde (C_6H_5CHO, melting point: $-26°C$, boiling point: $179°C$, density: 1.046) is a colorless liquid that is sometimes referred to as *artificial almond oil* or *oil of bitter almonds* because of the characteristic nutlike odor. Benzaldehyde is slightly soluble in water but is miscible in all proportions with alcohol or ether. On standing in air, benzaldehyde oxidizes readily to benzoic acid.

Commercially, benzaldehyde may be produced (1) by heating benzal chloride ($C_6H_5CHCl_2$) with calcium hydroxide:

$$C_6H_5CHCl_2 + Ca(OH)_2 \rightarrow C_6H_5CHO + CaO + 2HCl$$

(2) by heating calcium benzoate and calcium formate:

$$(C_6H_5COO)_2Ca + (HCOO)_2Ca \rightarrow 2C_6H_5CHO + 4\ 2CaCO_3$$

or (3) by boiling glucoside amygdalin of bitter almonds with a dilute acid.

Benzaldehyde is manufactured in two grades, technical and refined. The technical grade is largely used as an intermediate in the synthesis of other chemicals, such as benzyl benzoate, cinnamic aldehyde, and dyes.

Most of the technical grade is made by direct vapor-phase oxidation of toluene, although some is made by chlorinating toluene to benzal chloride, followed by alkaline or acid hydrolysis. For perfume and flavoring use, the refined, chlorine-free grade is required, which is economically produced by the direct vapor-phase oxidation of toluene with air at $500°C$.

$$C_6H_5CH_3 + [O] \rightarrow C_6H_5CH{=}O$$

It is claimed that a catalyst mixture of 93% uranium oxide and 7% molybdenum oxide gives relatively high yields. The oxidation is sometimes carried out in the liquid phase by using manganese dioxide/sulfuric acid at $40°C$.

Benzaldehyde is used as a flavoring material, in the production of cinnamic acid, in the manufacture of malachite green dye, as an ingredient in pharmaceuticals, and as an intermediate in chemical syntheses.

BENZENE

Benzene (C_6H_6, boiling point: 80°C, density: 0.8789, flash point: −11°C, ignition temperature: 538°C), is a volatile, colorless, and flammable liquid aromatic hydrocarbon possessing a distinct, characteristic odor. Benzene is practically insoluble in water (0.07 part in 100 parts at 22°C); and fully miscible with alcohol, ether, and numerous organic liquids.

For many years benzene (benzol) was made from coal tar, but new processes that consist of catalytic reforming of naphtha and hydrodealkylation of toluene are more appropriate. Benzene is a natural component of petroleum, but it cannot be separated from crude oil by simple distillation because of azeotrope formation with various other hydrocarbons. Recovery is more economical if the petroleum fraction is subjected to a thermal or catalytic process that increases the concentration of benzene.

Petroleum-derived benzene is commercially produced by reforming and separation, thermal or catalytic dealkylation of toluene, and disproportionation. Benzene is also obtained from pyrolysis gasoline formed in the steam cracking of olefins.

If benzene is the main product desired, a narrow light naphtha fraction boiling over the range 70 to 104°C is fed to the reformer, which contains a noble metal catalyst consisting of, for example, platinum-rhenium on a high-surface-area alumina support. The reformer operating conditions and type of feedstock determine the amount of benzene that can be produced. The benzene product is most often recovered from the reformate by solvent extraction techniques.

In the platforming process (Fig. 1), the feedstock is usually a straight-run, thermally cracked, catalytically cracked, or hydrocracked C_6 to 200°C naphtha. The feed is first hydrotreated to remove sulfur, nitrogen, or oxygen compounds that would foul the catalyst, and also to remove olefins present in cracked naphthas. The hydrotreated feed is then mixed with recycled hydrogen and preheated to 495 to 525°C at pressures of 116 to 725 psi (0.8 to 5 MPa). Typical hydrogen charge ratios of 4000 to 8000 standard cubic feet per barrel (scf/bbl) of feed are necessary.

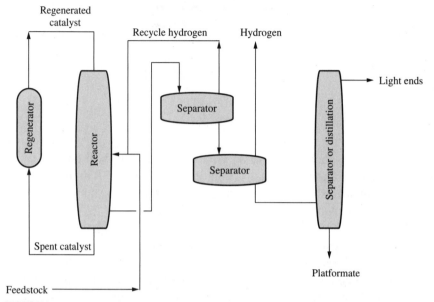

FIGURE 1   Benzene manufacture by the platforming process.

The feed is then passed through a stacked series of three or four reactors containing the catalyst (platinum chloride or rhenium chloride supported on silica or silica-alumina). The catalyst pellets are generally supported on a bed of ceramic spheres.

The product coming out of the reactor consists of excess hydrogen and a reformate rich in aromatics. The liquid product from the separator goes to a stabilizer where light hydrocarbons are removed and sent to a debutanizer. The debutanized platformate is then sent to a splitter where C_8 and C_9 aromatics are removed. The platformate splitter overhead, consisting of benzene, toluene, and nonaromatics, is then solvent extracted.

Solvents used to extract the benzene include tetramethylene sulfone (Fig. 2), diethylene glycol, N-methylpyrrolidinone process, dimethylformamide, liquid sulfur dioxide, and tetraethylene glycol.

Benzene is also produced by the hydrodemethylation of toluene under catalytic or thermal conditions.

In the catalytic hydrodealkylation of toluene (Fig. 3):

$$C_6H_5CH_3 + H_2 \rightarrow C_6H_6 + CH_4$$

toluene is mixed with a hydrogen stream and passed through a vessel packed with a catalyst, usually supported chromium or molybdenum oxides, platinum or platinum oxides, on silica or alumina. The operating temperatures range from 500 to 595°C and pressures are usually 580 to 870 psi

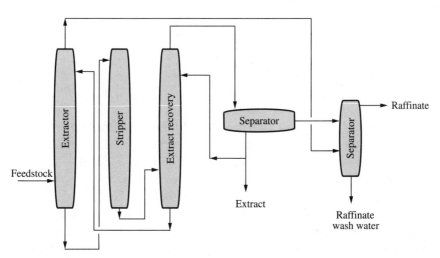

FIGURE 2   Benzene manufacture by sulfolane extraction.

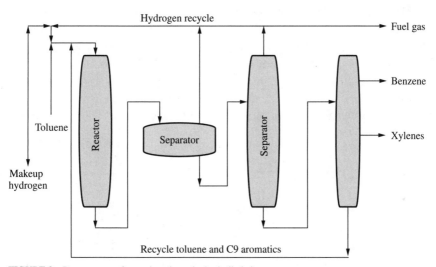

FIGURE 3   Benzene manufacture by toluene hydrodealkylation.

(4 to 6 MPa). The reaction is exothermic and temperature control (by injection of quench hydrogen) is necessary at several places along the reaction sequence. Conversions per pass typically reach 90 percent and selectivity to benzene is often greater than 95 percent. The catalytic process occurs at lower temperatures and offers higher selectivity but requires frequent regeneration of the catalyst. Products leaving the reactor pass through a separator where unreacted hydrogen is removed and recycled to the feed. Further fractionation separates methane from the benzene product.

Benzene is also produced by the transalkylation of toluene in which two molecules of toluene are converted into one molecule of benzene and one molecule of mixed xylene isomers.

In the process (Fig. 4), toluene and C_9 aromatics are mixed with liquid recycle and recycle hydrogen, heated to 350 to 530°C at 150 to 737 psi (1 to 5 MPa), and charged to a reactor containing a fixed bed of noble metal or rare earth catalyst with hydrogen-to-feedstock mole ratios of 5:1 to 12:1. Following removal of gases, the separator liquid is freed of light ends and the bottoms are then clay treated and fractionated to produce high-purity benzene and xylenes. The yield of benzene and xylene obtained from this procedure is about 92 percent of the theoretical.

Other sources of benzene include processes for steam cracking heavy naphtha or light hydrocarbons such as propane or butane to produce a liquid product (pyrolysis gasoline) rich in aromatics that contains up to about 65 percent aromatics, about 50 percent of which is benzene. Benzene can be recovered by solvent extraction and subsequent distillation.

Benzene can also be recovered from coal tar. The lowest-boiling fraction of the tar is extracted with caustic soda to remove tar acids, and the base oil is then distilled and further purified by hydrodealkylation.

Benzene is used as a chemical intermediate for the production of many important industrial compounds, such as styrene (polystyrene and synthetic rubber), phenol (phenolic resins), cyclohexane (nylon), aniline (dyes), alkylbenzenes (detergents), and chlorobenzenes. These intermedi-

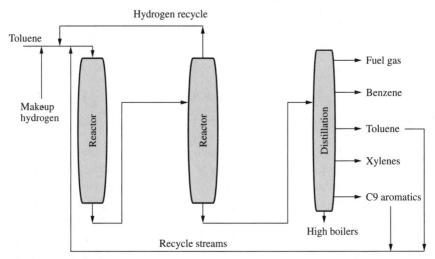

FIGURE 4   Benzene manufacture by the transalkylation of toluene.

ates, in turn, supply numerous sectors of the chemical industry producing pharmaceuticals, specialty chemicals, plastics, resins, dyes, and pesticides. In the past, benzene has been used in the shoe and garment industry as a solvent for natural rubber. Benzene has also found limited application in medicine for the treatment of certain blood disorders and in veterinary medicine as a disinfectant.

Benzene, along with other light high-octane aromatic hydrocarbons such as toluene and xylene, is used as a component of motor gasoline. Benzene is used in the manufacture of styrene, ethylbenzene, cumene, phenol, cyclohexane, nitrobenzene, and aniline. It is no longer used in appreciable quantity as a solvent because of the health hazards associated with it.

Ethylbenzene is made from ethylene and benzene and then dehydrogenated to styrene, which is polymerized for various plastics applications. Cumene is manufactured from propylene and benzene and then made into phenol and acetone. Cyclohexane, a starting material for some nylon, is made by hydrogenation of benzene. Nitration of benzene followed by reduction gives aniline, important in the manufacture of polyurethanes.

BENZINE

Benzine is a product of petroleum (boiling between 49 and 66°C) and is composed of aliphatic hydrocarbons.

Benzine should not be confused with benzene, which is a single chemical compound and an aromatic hydrocarbon.

BENZODIAZEPINES

Benzodiazepines are a series of compounds that have a benzene ring fused to a seven-membered ring containing two nitrogen atoms. The two most successful members of the group are diazepam (Valium®) and chlordiazepoxide (Librium®); Flurazepam (Dalmane®) is a hypnotic.

The production of diazepam (Fig. 1) starts from p-chloroaniline prepared from benzene by nitration, reduction of the nitro group to an amine, and chlorination of the o,p-directing aniline) followed by reaction with benzoyl chloride (from toluene by oxidation to benzoic acid, followed by acid chloride formation) in a Friedel-Crafts acylation. Acylation occurs at the o-position, and formation of the oxime derivative is followed by methylation and then acylation of the amino group with chloroacetyl chloride. Heating with base eliminates hydrogen chloride to form the ring and reduction of the amine oxide with hydrogen gives diazepam.

Diazepam is used for the control of anxiety and tension, the relief of muscle spasms, and the management of acute agitation during alcohol withdrawal, but it may be habit forming.

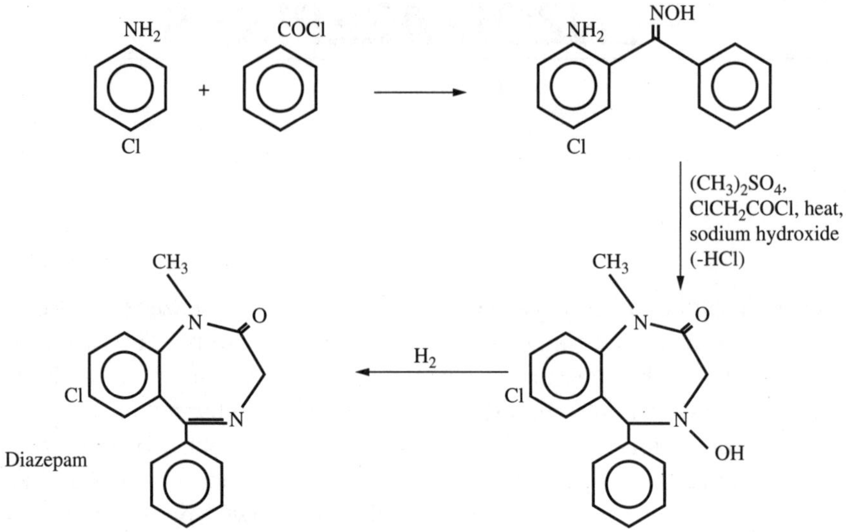

FIGURE 1   Manufacture of diazepam (Valium).

BENZOIC ACID

Benzoic acid (C_6H_5COOH, phenyl formic acid, melting point: 121.7°C, boiling point: 249.2°C, density: 1.266) is a white crystalline solid that sublimes readily at 100°C and is volatile in steam. Benzoic acid is insoluble in cold water but is readily soluble in hot water or in alcohol or ether.

Although benzoic acid occurs naturally in some substances, such as gum benzoin, dragon's blood resin, balsams, cranberries, and the urine of the ox and horse, the product is made on a large scale by synthesis from other materials. Benzoic acid can be manufactured by the liquid-phase oxidation of toluene by air in a continuous oxidation reactor operated at moderate pressure and temperature:

$$2C_6H_5CH_3 + 3O_2 \rightarrow 2C_6H_5COOH + 4H_2O$$

Benzoic acid also can be obtained as a by-product of the manufacture of benzaldehyde from benzal chloride or benzyl chloride.

Benzoic acid is used as a starting or intermediate material in various industrial organic syntheses, especially in the manufacture of terephthalic acid. Benzoic acid forms benzoates; e.g., sodium benzoate and calcium benzoate which, when heated with calcium oxide, yields benzene and calcium. With phosphorus trichloride, benzoic acid forms benzoyl chloride (C_6H_5COCl), an agent for the transfer of the benzoyl group (C_6H_5CO-). Benzoic acid reacts with chlorine to form *m*-chlorobenzoic acid and reacts with nitric acid to form *m*-nitrobenzoic acid. Benzoic acid forms a number of useful esters such as methyl benzoate, ethyl benzoate, glycol dibenzoate, and glyceryl tribenzoate.

BENZYL ACETATE

Benzyl acetate ($C_6H_5CH_2OCOCH_3$) is a widely used ester because of its floral odor.

It is prepared by esterification of benzyl alcohol, by heating with either an excess of acetic anhydride or acetic acid with mineral acids.

$$C_6H_5CH_2OH + CH_3CO_2H \rightarrow C_6H_5CH_2OCOCH_3$$

The product is purified by treatment with boric acid and distilled to a purity in excess of 98%.

BENZYL ALCOHOL

Benzyl alcohol is employed in pharmaceuticals and lacquers. This alcohol has a much weaker odor than its esters.

Benzyl alcohol is manufactured by hydrolysis of benzyl chloride.

$$C_6H_5CH_2Cl \rightarrow C_6H_5CH_2OH$$

See *Benzyl Acetate*

BISPHENOL A

Bisphenol A is made by reacting phenol with acetone in the presence of an acid catalyst (Fig. 1). The temperature of the reaction is maintained at 50°C (120°F) for about 8 to 12 hours. A slurry of bisphenol A is formed, which is neutralized and distilled to remove excess phenol.

$$2C_6H_5OH + CH_3COCH_3 \rightarrow p\text{--}HOC_6H_4C(CH_3)_2C_6H_4OH$$

Bisphenol A is used in the production of epoxy resins and polycarbonate resins.

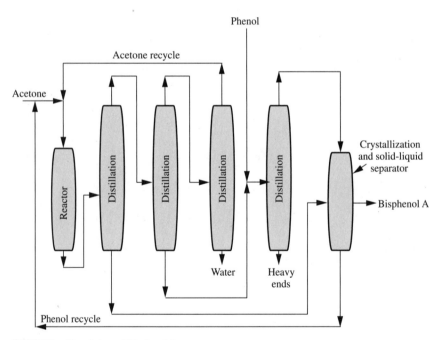

FIGURE 1 Manufacture of Bisphenol A.

BORAX

Borax is a hydrated sodium borate mineral ($Na_2B_4O_7 \cdot 10H_2O$, density: 1.715). It crystallizes in the monoclinic system, usually in short prismatic crystals. Borax ranges from colorless through gray, blue to greenish and has a vitreous to resinous luster of translucent to opaque character. Borax is a product of evaporation from shallow lakes.

Borax is used in antiseptics and medicines, as a flux in smelting, soldering, and welding operations; as a deoxidizer in nonferrous metals; as a neutron absorber for atomic energy shields; in rocket fuels; and as extremely hard, abrasive boron carbide (harder than corundum).

BORON COMPOUNDS

The important naturally occurring ores of boron are colemanite ($Ca_2B_6O_{11}$ $5H_2O$), tincal ($Na_2B_4O_7 \cdot 10H_2O$), and boracite ($2Mg_3B_8O_{15} \cdot MgCl_2$). Boron-containing brines and kernite, or rasorite ($Na_2B_4O_7\ 4H_2O$), are also sources of boron compounds.

Crude and refined hydrated sodium borates and hydrous boric acid are produced from kernite and tincal. The ore is fed to the dissolving plant and mixed with hot recycle liquor. Liquor and fine insolubles are fed to a primary thickener. The strong liquors are crystallized in a continuous vacuum crystallizer.

Another process involves the use of an organic solvent to extract the borax brines. Boric acid is extracted with kerosene, carrying a chelating agent. In a second mixer-settler system, dilute sulfuric acid strips the borates from the chelate. The aqueous phase with boric acid, potassium sulfate, and sodium sulfate is purified by carbon treatment and evaporated in two evaporator-crystallizers. From the first, pure boric acid is separated, and from the other, a mixture of sodium and potassium sulfates.

Borax (tincal, $Na_2B_4O_7 \cdot 10H_2O$), is the most important industrial compound of boron. The largest single use is in the manufacture of glass-fiber insulation. Boric acid (H_3BO_4) is a weak acid that is used in the manufacture of glazes and enamels for pottery. Its main uses are as a fire retardant for cellulosic insulation and in the manufacture of borosilicate glasses and textile-grade glass fibers.

BROMAL

*See **Chloral.***

BROMINE

Bromine (freezing point: $-7.3°C$, boiling point: $58.8°C$, density: 3.1226) is a member of the halogen family and is a heavy, dark-red liquid.

Bromine is produced from seawater, in which bromine occurs in concentrations of 60 to 70 ppm, and from natural brine, where the concentration of bromine may be as high as 1300 ppm. It can also be produced from waste liquors resulting from the extraction of potash salts from carnallite deposits.

Bromine is isolated from sea water by air-blowing it out of chlorinated seawater.

$$2NaBr + Cl_2 \rightarrow 2NaCl + Br_2$$

In ocean water, where the concentration of bromine is relatively dilute, air has proved to be the most economical blowing-out agent. However, in the treatment of relatively rich bromine sources such as brines, *steaming out* the bromine vapor is more satisfactory.

The steaming-out process (Fig. 1) process involves preheating the brine to 90°C in a heat exchanger and passing it down a chlorinator tower. After partial chlorination, the brine flows into a steaming-out tower, where steam is injected at the bottom and the remaining chlorine is introduced. The halogen-containing vapor is condensed and gravity separated. The top water-halogen layer is returned to the steaming-out tower, and the crude halogen (predominantly bromine) bottom layer is separated and purified. Crude bromine is purified by redistillation or by passing the vapors over iron filings that remove any chlorine impurity.

Bromine is used for the production of alkali bromides that cannot be manufactured by the action of caustic soda on bromine because hypobromites and bromates are also produced. Thus, the van der Meulen process from the production of potassium bromide involves treating bromine with potassium carbonate in the presence of ammonia.

$$K_2CO_3 + 3Br_2 + 2NH_3 \rightarrow 6KBr + N_2 + 3CO_2 + 3H_2O$$

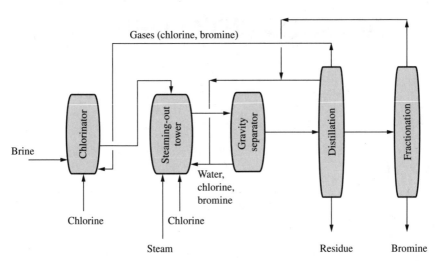

FIGURE 1   Manufacture of bromine from brine.

BROMOACETALDEHYDE

*See **Chloral.***

BTX AROMATICS

BTX (benzene-toluene-xylene) mixtures are an important petroleum refinery stream that is separated by extractive distillation (Fig. 1) from a hydrocarbon stream, usually a reformate, and followed by downstream fractionation for isolation of the pure materials for further treatment and use (Fig. 2).

The BTX stream provides the individual components of the mixture for further use (Fig. 3).

See Benzene, Toluene, Xylenes.

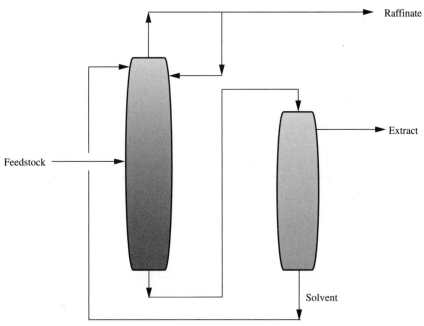

FIGURE 1 Schematic of extractive distillation.

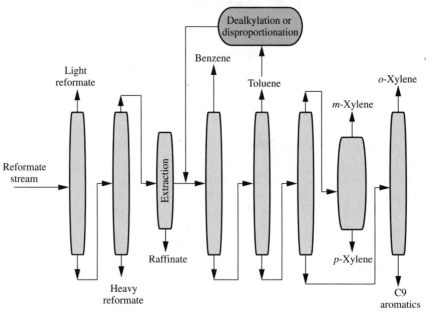

FIGURE 2   Separation of a BTX stream into its constituents.

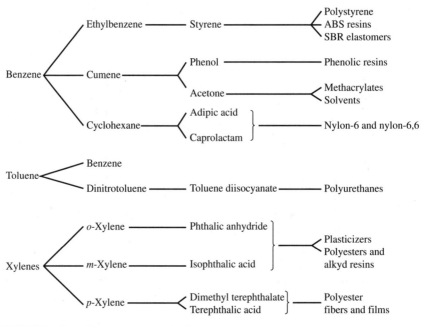

FIGURE 3   Uses of benzene, toluene, and xylenes.

BUTADIENE

Butadiene (1,3-butadiene, boiling point: –4.4°C, density: 0.6211, flash point: –85°C) is made by steam cracking and by the dehydrogenation of butane or the butenes using an iron oxide (Fe_2O_3) catalyst.

$$CH_3CH_2CH_2CH_3 \rightarrow CH_2=CHCH=CH_2 + 2H_2$$

$$CH_3CH_2CH=CH_2 \rightarrow CH_2=CHCH=CH_2 + H_2$$

$$CH_3CH=CHCH_3 \rightarrow CH_2=CHCH=CH_2 + H_2$$

In the process (Fig. 1), the crude C_4 fraction is extracted with acetone, furfural, or other solvents to remove alkanes such as *n*-butane, *iso*-butane, and small amounts of pentanes, leaving only 1- and 2-butenes and iso-butene. The isobutene is removed by extraction with sulfuric acid because it oligomerizes more easily.

The straight-chain 1- and 2-butenes are preheated to 600°C in a furnace, mixed with steam as a diluent to minimize carbon formation, and passed through a 5-m-diameter reactor with a bed of iron oxide pellets (or calcium nickel phosphate) 90 to 120 cm deep (contact time 0.2 second) at 620 to 750°C. The material is cooled and purified by *fractional distillation* and extraction with solvents such as furfural, acetonitrile, dimethylformamide (DMF), and *N*-methylpyrollidone (NMP) (Fig. 2).

Extractive distillation (Fig. 3) is used where the C_4 compounds other than butadiene are distilled while the butadiene is complexed with the solvent. The solvent and butadiene pass from the bottom of the column and are then separated by distillation.

Butadiene is a colorless, odorless, flammable gas, with a boiling point of –4.7°C and is used for the manufacture of polybutadiene, nitrile rubber, chloroprene, and various other polymers. An important synthetic elastomer is styrene-butadiene rubber (SBR) in the automobile tire industry. Specialty elastomers are polychloroprene and nitrile rubber, and an important plastic is acrylonitrile/butadiene/styrene (ABS) terpolymer. Butadiene is made into adiponitrile, which is converted into hexamethylenediamine (HMDA), one of the monomers for nylon.

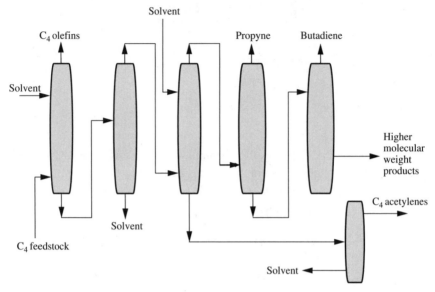

FIGURE 1   Butadiene manufacture from a refinery C_4 stream.

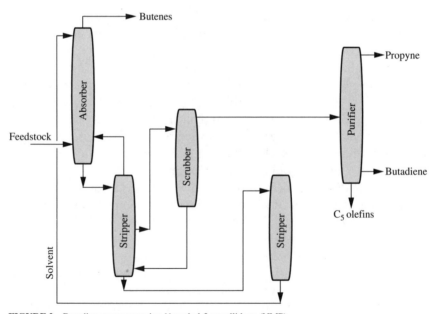

FIGURE 2   Butadiene recovery using *N*-methyl-2-pyrollidone (NMP).

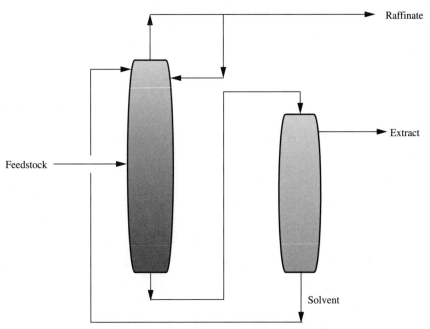

FIGURE 3 Schematic of extractive distillation.

BUTANE

See Liquefied Petroleum Gas.

BUTANEDIOL

1,4-butanediol (tetramethylene glycol, 1,4-butylene glycol; melting point: 20.2°C, boiling point: 228°C, density:1.017, flash point: 121°C) was first prepared in 1890 by acid hydrolysis of N,N'-dinitro-1,4-butanediamine. Other early preparations were by reduction of succinaldehyde or succinic esters and by saponification of the diacetate prepared from 1,4-dihalobutanes. Catalytic hydrogenation of butynediol, now the principal commercial route, was first described in 1910.

Butanediol is manufactured by way of hydrogenation of butynediol (the Reppe process):

$$HC \equiv CH + 2HCHO \rightarrow HOCH_2C \equiv CCH_2OH \rightarrow$$
$$HOCH_2CH_2CH_2CH_2OH$$

An alternative route involving acetoxylation of butadiene and has come on stream, and, more recently, a route based upon hydroformylation of allyl alcohol has also been used. Another process, involving chlorination of butadiene, hydrolysis of the dichlorobutene, and hydrogenation of the resulting butenediol, has been practiced.

A more modern process involves the use of maleic anhydride as the starting material. In the process (Fig. 1), maleic anhydride is first esterified with methanol and the ester is fed to a low-pressure vapor-phase hydrogenation system where it is converted to butanediol.

Butanediol is specified as 99.5% minimum pure, determined by gas chromatography, solidifying at 19.6°C minimum. Moisture is 0.04% maximum, determined by Karl Fischer analysis (directly or by a toluene azeotrope).

Butanediol is much less toxic than its unsaturated analogs and it is neither a primary skin irritant nor a sensitizer. Because of its low vapor pressure, there is ordinarily no inhalation problem but, as with all chemicals, unnecessary exposure should be avoided.

The uses of butanediol are determined by the chemistry of the two primary hydroxyls. Esterification is normal and it is advisable to use nonacidic

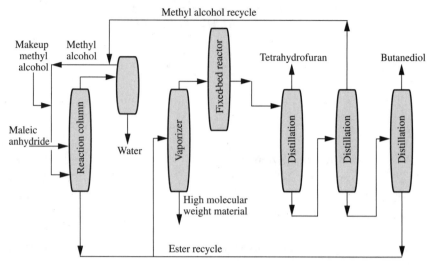

FIGURE 1   Manufacture of 1,4-butanediol from maleic anhydride.

catalysts for esterification and transesterification to avoid cyclic dehydration. When carbonate esters are prepared at high dilutions, some cyclic ester is formed; more concentrated solutions give a polymeric product.

Ethers are formed in the usual way; the *bis*-chloromethyl ether is obtained by using formaldehyde and hydrogen chloride. With aldehydes or their derivatives, butanediol forms acetals, either 7-membered rings (1,3-dioxepanes) or linear polyacetals; the rings and chains are easily intraconverted.

Thionyl chloride readily converts butanediol to 1,4-dichlorobutane and hydrogen bromide gives 1,4-dibromobutane. A procedure using 48% hydrobromic acid with a Dean-Stark water trap gives good yields of 4-bromobutanol, free of diol and dibromo compound.

With various catalysts, butanediol adds carbon monoxide to form adipic acid. Heating with acidic catalysts dehydrates butanediol to tetrahydrofuran. With dehydrogenation catalysts, such as copper chromite, butanediol forms butyrolactone. With certain cobalt catalysts, both dehydration and dehydrogenation occur, giving 2,3-dihydrofuran.

Heating butanediol or tetrahydrofuran with ammonia or an amine in the presence of an acidic heterogeneous catalyst gives pyrrolidines. With a dehydrogenation catalyst, amino groups replace one or both of the hydroxyl groups.

Vapor-phase oxidation over a promoted vanadium pentoxide catalyst gives a 90 percent yield of maleic anhydride. Liquid-phase oxidation with a supported palladium catalyst gives 55 percent of succinic acid.

The largest uses of butanediol are internal consumption in manufacture of tetrahydrofuran and butyrolactone. The largest merchant uses are for poly(butylene terephthalate) resins and in polyurethanes, both as a chain extender and as an ingredient in a hydroxyl-terminated polyester used as a macroglycol. Butanediol is also used as a solvent, as a monomer for various condensation polymers, and as an intermediate in the manufacture of other chemicals.

ISO-BUTANE

Iso-butane [(CH$_3$)$_3$CH] can be isolated from the petroleum C$_4$ fraction or from natural gas by extraction and distillation.

There are two major uses of *iso*-butane. One is dehydrogenation to isobutylene followed by conversion of the isobutylene to the gasoline additive methyl *t*-butyl ether (MTBE). However, current environmental issues may ban this gasoline additive.

Iso-butane is also oxidized to the hydroperoxide and then reacted with propylene to give propylene oxide and *t*-butyl alcohol. The *t*-butyl alcohol can be used as a gasoline additive, or dehydrate to *iso*-butylene.

BUTENE-1

The steam cracking of naphtha and catalytic cracking in the refinery produce the C_4 stream, which includes butane, 1-butene (butylene), *cis-* and *trans-*2-butene, isobutylene, and butadiene.

$$C_nH_{2n+2} \rightarrow CH_3CH_2CH_2CH=CH_2$$

Extracting the isobutylene with sulfuric acid and distilling the 1-butene away from butane and butadiene is used for separation of the 1-butene. It is also made by Ziegler (or non-Ziegler) oligomerization of ethylene (Fig. 1).

1-butene is used as a comonomer to make polyethylene and is used to make valeraldehyde (pentanal) by the oxo process, polybutene-1, and butylene oxide.

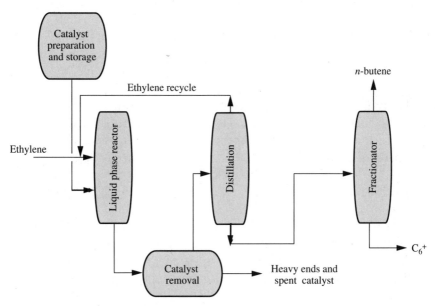

FIGURE 1 Manufacture of butene-1 (*n*-butene) by dimerization of ethylene.

BUTENEDIOL

2-butene-1,4-diol (melting point: 11.8°C, boiling point: 234°C, density: 1.070, flash point: 128°C) is the only commercially available olefinic diol with primary hydroxyl groups. The commercial product consists almost entirely of the *cis* isomer. Butenediol is very soluble in water, lower alcohols, and acetone, but is almost insoluble in aliphatic or aromatic hydrocarbons.

Butenediol is manufactured by partial hydrogenation of butynediol. Although suitable conditions can lead to either *cis* or *trans* isomers, the commercial product contains almost exclusively *cis*-2-butene-1,4-diol. The *trans* isomer, available at one time by hydrolysis of 1,4-dichloro-2-butene, is unsuitable for the major uses of butenediol involving Diels-Alder reactions. The liquid-phase heat of hydrogenation of butynediol to butenediol is 156 kJ/mol (37.28 kcal/mol).

The original German process used either carbonyl iron or electrolytic iron as hydrogenation catalyst. The fixed-bed reactor was maintained at 50 to 100°C and 20.26 MPa (200 atm) of hydrogen pressure, giving a product containing substantial amounts of both butynediol and butanediol. Newer, more selective processes use more active catalysts at lower pressures. In particular, supported palladium, alone or with promoters, has been found useful.

Purity is determined by gas chromatography. Technical grade butenediol, specified at 95% minimum, is typically 96 to 98% butenediol. The *cis* isomer is the predominant constituent; 2 to 4% is *trans*. Principal impurities are butynediol (specified as 2.0% maximum, typically less than 1%), butanediol, and the 4-hydroxybutyraldehyde acetal of butenediol. Moisture is specified at 0.75% maximum (Karl-Fischer titration). Technical-grade butenediol freezes at about 8°C.

Butenediol is noncorrosive and stable under normal handling conditions. It is a primary skin irritant but not a sensitizer; contact with skin and eyes should be avoided. It is much less toxic than butynediol. The LD_{50} is 1.25 mL/kg for white rats and 1.5 mL/kg for guinea pigs.

Butenediol is used to manufacture the insecticide Endosulfan, other agricultural chemicals, and pyridoxine (vitamin B_6). Small amounts are consumed as a diol by the polymer industry.

In addition to the usual reactions of primary hydroxyl groups and of double bonds, cis-butenediol undergoes a number of cyclization reactions.

The hydroxyl groups can be esterified normally: the interesting diacrylate monomer and the biologically active haloacetates have been prepared in this manner. Reactions with dibasic acids have given polymers capable of being cross-linked or suitable for use as soft segments in polyurethanes. Polycarbamic esters are obtained by treatment with a diisocyanate or via the bischloroformate.

The hydroxyl groups can be alkylated in the usual manner. Hydroxyalkyl ethers may be prepared with alkylene oxides and chloromethyl ethers by reaction with formaldehyde and hydrogen chloride. The terminal chlorides can be easily converted to additional ether groups.

ISO-BUTENE

Iso-butylene is made largely by the catalytic and thermal cracking of hydrocarbons. Other four-carbon (C_4) products formed in the reaction include other butenes, butane, *iso*-butane, and traces of butadiene. The most widely used process is the fluid catalytic cracking of gas oil; other methods are delayed coking and flexicoking.

$$C_nH_{2n+2} \rightarrow (CH_3)_2C{=}CH_2$$

Iso-butene is used as a raw material in the production of methyl *t*-butyl ether (MTBE, gasoline additive) and butylated hydroxytoluene (an antioxidant) as well as for the production of alkylate for gasoline.

N-BUTENE

n-Butylene (butene-1) is manufactured by the dimerization of ethylene obtained from the catalytic and thermal cracking of hydrocarbons (Fig. 1).

$$2CH_2{=}CH_2 \rightarrow CH_3CH_2CH{=}CH_2$$

n-Butene thus produced is of suitably high purity to be used for various polymerization and copolymerization processes.

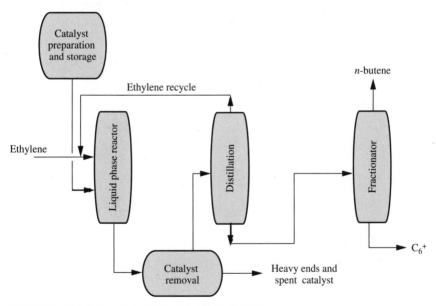

FIGURE 1   Manufacture of *n*-butene by dimerization of ethylene.

BUTYL ACRYLATE

Acrylates are still produced by a modified Reppe process that involves the reaction of acetylene, the appropriate alcohol (in the case of butyl acrylate, butyl alcohol is used), and carbon monoxide in the presence of an acid. The process is continuous, and a small amount of acrylates is made this way.

$$CH{\equiv}CH + ROH + CO \rightarrow CH_2{=}CHCOOR$$

The most economical method of acrylate production is that of the direct oxidation of propylene to acrylic acid, followed by esterification.

$$CH_2{=}CHCH_3 + O_2 \rightarrow CH_2{=}CHCOOH$$

$$CH_2{=}CHCOOH + ROH \rightarrow CH_2{=}CHCOOR + H_2O$$

Esters of acrylic acid (acrylates) are used for manufacture of coatings, textiles, fibers, polishes, paper, and leather.

ISO-BUTYL ALCOHOL

Iso-butyl alcohol (boiling point: 108.1°C, density: 0.8018) is of particular interest in motor fuels because of its high octane rating, which makes it desirable as a gasoline-blending agent.

This alcohol can be reacted with methanol in the presence of a catalyst to produce methyl-*t*-butyl ether. Although it is currently cheaper to make *iso*-butyl alcohol from *iso*-butene (*iso*-butylene), it can be synthesized from syngas with alkali-promoted zinc oxide catalysts at temperatures above 400°C (750°F).

See *Iso-Butene, Oxo Reaction, Synthesis Gas.*

n-BUTYL ALCOHOL

n-Butanol (*n*-butanol; boiling point: 117.7°C, density: 0.8097, flash point: 28.9°C) can be obtained from carbohydrates (such as molasses and grain) by fermentation. Acetone and ethanol are produced as by-products.

$$C_6H_{12}O_6 \rightarrow CH_3CH_2CH_2CH_2OH + CH_3COCH_3$$
$$+ CH_3CH_2OH + CO_2 + H_2$$

Propylene and synthesis gas give *n*-butyl alcohol; *iso*-butyl alcohol is a by-product.

$$CH_3CH=CH_2 + CO_2 + H_2 \rightarrow CH_3CH_2CH_2CHO$$

$$CH_3CH_2CH_2CHO + H_2 \rightarrow CH_3CH_2CH_2CH_2OH$$

n-Butyl alcohol is used in the manufacture of butyl acrylate and methacrylate, glycol ethers, solvents, butyl acetate, and plasticizers.

*See **Oxo Reaction, Synthesis Gas.***

t-BUTYL ALCOHOL

t-Butyl alcohol (melting point: 25.8°C, boiling point: 82.4°C, density: 0.7866, flash point: 11.1°C) is a low-melting solid that, after melting, exists as a colorless liquid.

t-Butyl alcohol is produced by the hydration of *iso*-butene. The favored tertiary carbocation intermediate limits the possible alcohols produced to only this one.

$$(CH_3)_2C{=}CH_2 \rightarrow (CH_3)_3COH$$

See *Iso-Butene*.

BUTYL VINYL ETHER

See Vinyl Ethers.

BUTYNEDIOL

Butynediol (2-butyne-1,4-diol; melting point: 58°C, boiling point: 248°C, density: 1.114, flash point: 152°C) is a stable crystalline solid but violent reactions can take place in the presence of certain contaminants, particularly at elevated temperatures. In the presence of certain heavy-metal salts, such as mercuric chloride, dry butynediol can decompose violently. Heating with strongly alkaline materials should be avoided.

Butynediol was first synthesized in 1906 by reaction of acetylene *bis*(magnesium bromide) with paraformaldehyde.

$$HC{\equiv}CH + 2HCH{=}O \rightarrow HOCH_2C{\equiv}CCH_2OH$$

All manufacturers of butynediol use this formaldehyde ethynylation process, and yields of butynediol may be in excess of 90 percent, in addition to 4 to 5% propargyl alcohol.

$$HC{\equiv}CH + HCH{=}O \rightarrow HC{\equiv}CH_2OH$$

Most butynediol produced is consumed in the manufacture of butanediol and butenediol. Butynediol is also used for conversion to ethers with ethylene oxide and in the manufacture of brominated derivatives that are useful as flame retardants. Butynediol was formerly used in a wild oat herbicide, Carbyne (Barban), 4-chloro-2-butynyl-*N*-(3-chlorophenyl)carbamate ($C_{11}H_9Cl_2NO_2$).

Butynediol undergoes the usual reactions of primary alcohols that contribute to its use as a chemical intermediate. Because of its rigid, linear structure, many reactions forming cyclic products from butanediol or *cis*-butenediol give only polymers with butynediol. Both hydroxyl groups can be esterified normally, and the monoesters are readily prepared as mixtures with diesters and unesterified butynediol, but care must be taken in separating them because the monoesters disproportionate easily.

The hydroxyl groups can be alkylated with the alkylating agents, although a reverse treatment is used to obtain aryl ethers; for example, treatment of butynediol toluene sulfonate or dibromobutyne with a phenol gives the corresponding ether. Reactions of butynediol with alkylene oxides give ether alcohols.

In the presence of acid catalysts, butynediol and aldehydes or acetals give polymeric acetals, useful intermediates for acetylenic polyurethanes suitable for high-energy solid propellants.

$$HOCH_2C{\equiv}CCH_2OH \rightarrow HO(CH_2C{\equiv}CCH_2OCH_2O)_nH$$

Electrolytic oxidation gives acetylene dicarboxylic acid (2-butyne-dioic acid) in good yields. Butynediol can be hydrogenated partway to butenediol ($HOCH_2CH{=}CHCH_2OH$) or completely to butanediol ($HOCH_2CH_2CH_2CH_2OH$).

Addition of halogens proceeds stepwise, sometimes accompanied by oxidation. Iodine forms 2,3-diiodo-2-butene-1,4-diol and, depending on conditions, bromine gives 2,3-dibromo-2-butene-l, 4-diol, 2,2,3,3-tetrabromobutane-1, 4-diol, mucobromic acid, or 2-hydroxy-3,3,4,4-tetrabromotetrahydrofuran. Addition of chlorine is attended by more oxidation, which can be lessened by esterification of the hydroxyl groups.

Butynediol is more difficult to polymerize than propargyl alcohol, but it cyclotrimerizes to hexamethylolbenzene (benzenehexamethanol) with a nickel carbonyl-phosphine catalyst or a rhodium chloride–arsine catalyst.

ISO-BUTYRALDEHYDE

See n-Butyraldehyde.

n-BUTYRALDEHYDE

n-butyraldehyde (boiling point: 74.8°C, density: 0.8016) is made by the reaction of propylene (propene-1, $CH_3CH=CH_2$), carbon monoxide, and hydrogen (synthesis gas) at 130 to 175°C and 3675 psi (25.3 MPa) over a rhodium carbonyl, cobalt carbonyl, or ruthenium carbonyl catalyst (Fig. 1).

$$CH_3CH=CH_2 + CO + H_2 \rightarrow CH_3CH_2CH_2CH=O$$

The reaction is referred to as the *oxo process*, and a second product of the reaction is *iso*-butyraldehyde (boiling point: 64.1°C, density: 0.7891).

$$CH_3CH=CH_2 + CO + H2 \rightarrow CH_3CH(CH_3)CH=O$$

The ratio of the *normal* to the *iso* product is approximately 4:1. Other catalysts produce different *normal*-to-*iso* product ratios under different conditions. For example, a rhodium catalyst can be used at lower temperatures and pressures and gives a *normal*-to-*iso* product ratio of approximately 16:1.

Butyraldehyde is used for the production of *n*-butyl alcohol.

$$CH_3CH_2CH_2CH=O + H_2 \rightarrow CH_3CH_2CH_2CH_2OH$$

See Oxo Reaction, Synthesis Gas.

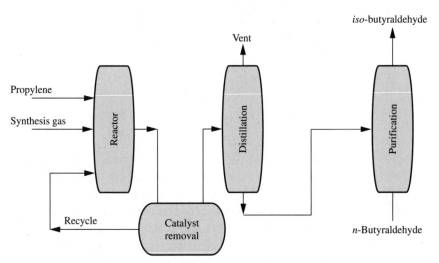

FIGURE 1   Manufacture of butyraldehyde from propylene and synthesis gas.

BUTYROLACTONE

γ-butyrolactone (dihydro-2,3H-furanone, boiling point: 204°C, density: 1.129, flash point: 98°C) was first synthesized in 1884 by an internal esterification of 4-hydroxybutyric acid. The principal commercial source of this material is dehydrogenation of butanediol. The manufacture of butyrolactone by hydrogenation of maleic anhydride is also a viable route.

Butyrolactone is completely miscible with water and most organic solvents. It is only slightly soluble in aliphatic hydrocarbons. It is a good solvent for many gases, for most organic compounds, and for a wide variety of polymers.

Butyrolactone is used to produce N-methyl-2-pyrrolidinone and 2-pyrrolidinone, by reaction with methylamine or ammonia, respectively. Considerable amounts are used as a solvent for agricultural chemicals and polymers, in dyeing and printing, and as an intermediate for various chemical syntheses.

Butyrolactone undergoes the reactions typical of γ-lactones. Particularly characteristic are ring openings and reactions in which ring oxygen is replaced by another heteroatom. There is also marked reactivity of the hydrogen atoms alpha to the carbonyl group.

With acid catalysts, butyrolactone reacts with alcohols rapidly even at room temperature to produce esters of 4-hydroxybutyric acid. The esters can be separated by a quick flash distillation at high vacuum.

Butyrolactone reacts rapidly and reversibly with ammonia or an amine forming 4-hydroxybutyramides, which dissociate to the starting materials when heated. At high temperatures and pressures, the hydroxybutyramides slowly and irreversibly dehydrate to pyrrolidinones; this dehydration is accelerated by use of a copper-exchanged Y-zeolite or magnesium silicate.

CAFFEINE, THEOBROMINE, AND THEOPHYLLINE

Caffeine (melting point: 238°C, sublimes at 178°C; density: 1.23), theobromine, and theophylline are xanthine derivatives classified as central nervous stimulants, but differing markedly in their properties. They can be extracted from a number of natural sources.

Caffeine has been isolated from waste tea and from the decaffeinization of coffee by extraction at 70°C, using rotating countercurrent drums and an organic solvent, frequently trichloroethylene. The solvent is drained off, and the beans steamed to remove residual solvent. The extraction solvent is evaporated, and the caffeine is hot-water-extracted from the wax, decolorized with carbon, and recrystallized.

Caffeine is also manufactured synthetically by, for instance, the methylation of theobromine and also total synthesis by methylation and other reactions based upon urea.

Caffeine is used by the pharmaceutical industry and also by the soft-drink industry for cola-style drinks.

CALCITE

*See **Calcium Carbonate.***

CALCIUM ACETATE

Calcium acetate is manufactured by the reaction of calcium carbonate or calcium hydroxide with acetic acid.

$$Ca(OH)_2 + CH_3CO_2H \rightarrow Ca(OCOCH_3)_2 + 2H_2O$$

Calcium acetate was formerly used to manufacture acetone by thermal decomposition:

$$Ca(OCOCH_3)_2 \rightarrow (CH_3)_2C{=}O + CaO + CO_2$$

but now it is employed largely in the dyeing of textiles.

*See **Acetone**.*

CALCIUM ARSENATE

Calcium arsenate is produced by the reaction of calcium chloride, calcium hydroxide, and sodium arsenate or lime and arsenic acid:

$$2CaCl_2 + Ca(OH)_2 + 2Na_2HAsO_4 \rightarrow Ca_3(AsO_4)_2 + 4NaCl + 2H_2O$$

Some free lime (CaO) is usually present. Calcium arsenate is used extensively as an insecticide and as a fungicide.

CALCIUM BROMIDE

Calcium bromide ($CaBr_2$) and calcium iodide (CaI_2) have properties similar to those of calcium chloride ($CaCl_2$). They are prepared by the action of the halogen acids (HX) on calcium oxide or calcium carbonate.

$$CaO + HX \rightarrow CaX_2 + H_2O$$

where X is bromine (Br) or iodine (I).

Calcium bromide and calcium iodide are sold as hexahydrates for use in medicine and photography. Calcium fluoride (CaF_2) occurs naturally as fluorspar.

See **Fluorine.**

CALCIUM CARBONATE

Calcium carbonate ($CaCO_3$) occurs naturally as calcite (density: 2.7), a widely distributed mineral. Calcite is a common constituent of sedimentary rocks, as a vein mineral, and as deposits from hot springs and in caves as stalactites and stalagmites. Calcite is white or colorless through shades of gray, red, yellow, green, blue, violet, brown, or even black when charged with impurities; streaked, white; transparent to opaque. It may occasionally show phosphorescence or fluorescence.

Calcium carbonate is one of several important inorganic chemicals (Fig. 1) and is widely used in both its pure and its impure states. As marble chips, it is sold in many sizes as a filler for artificial stone, for the neutralization of acids,

Starting material	Reactant or process	Primary product	Reactant or process	Secondary products
Sulfur	Contact process	Sulfuric acid		Aluminum sulfate
			Wet process	Phosphoric acid
				Ammonium sulfate
Air	Liquefaction	Liquid nitrogen		Ammonia
				Nitric acid
				Ammonium nitrate
	Liquefaction	Liquid oxygen		
Methane	Steam reforming	Synthesis gas	Haber process	Ammonia
				Nitric acid
				Ammonium nitrate
				Urea
Calcium carbonate	Calcining	Carbon dioxide		
		Calcium oxide	Water	Calcium hydroxide
Sodium chloride	Solvay process	Calcium chloride		
	Solvay process	Sodium carbonate		Sodium silicate
				Silica gel
	Electrolysis	Sodium hydroxide		
	Electrolysis	Chlorine		Chlorinated hydrocarbons
				Hydrogen chloride

FIGURE 1 Important inorganic chemicals.

and for chicken grit. Marble dust is employed in abrasives and in soaps. Crude, pulverized limestone is used in agriculture to *sweeten* soils, and pulverized arid levigated limestone is used to replace imported chalk and whiting.

Whiting is pure, finely divided calcium carbonate prepared by wet grinding and levigating natural chalk. Whiting mixed with 18% boiled linseed oil yields *putty*, which sets by oxidation and by formation of the calcium salt. Much whiting is consumed in the ceramic industry. Precipitated, or artificial, whiting arises through precipitation, e.g., from reacting a boiling solution of calcium chloride with a boiling solution of sodium carbonate or passing carbon dioxide into a milk-of-lime suspension. Most of the latter form is used in the paint, rubber, pharmaceutical, and paper industries.

CALCIUM CHLORIDE

Calcium chloride is obtained from natural brine that typically contains 14% sodium chloride (NaCl), 9% calcium chloride ($CaCl_2$), and 3% magnesium chloride ($MgCl_2$). Evaporation precipitates sodium chloride, and magnesium chloride is removed by adding slaked lime to precipitate magnesium hydroxide.

$$MgCl_2 + Ca(OH)_2 \rightarrow Mg(OH)_2 + CaCl_2$$

Calcium chloride is used on roads for dust control in the summer and deicing in the winter, since it is less corrosive to concrete than is sodium chloride. Other applications of calcium chloride include thawing coal, as a component of oil and gas well fluids, and as antifreeze for concrete.

CALCIUM FLUORIDE

See Calcium Bromide, Fluorine.

CALCIUM HYPOCHLORITE

Calcium hypochlorite is manufactured by chlorination of calcium hydroxide [CaOH)$_2$] followed by the separation of the calcium hypochlorite [Ca(OCl)$_2$] through salting out from solution with sodium chloride.

$$2Ca(OH)_2 + 2Cl_2 \rightarrow Ca(OCl)_2 + CaCl_2 + H_2O$$

Calcium hypochlorite is also manufactured by the formation under refrigeration of the complex salt Ca(OCl)$_2$ NaOCl·NaCl·12H$_2$O, which is prepared by the chlorination of a mixture of sodium hydroxide and calcium hydroxide. The salt is reacted with a chlorinated lime slurry, filtered to remove salt, and dried, resulting finally in a stable product containing 65 to 70% calcium hypochlorite.

The great advantage of calcium hypochlorite is that it does not decompose on standing as does bleaching powder. It is also twice as strong as ordinary bleaching powder and is not hygroscopic.

CALCIUM IODIDE

See **Calcium Bromide.**

CALCIUM LACTATE

Calcium lactate is manufactured by the reaction of calcium carbonate ($CaCO_3$) or hydroxide [$Ca(OH)_2$] with lactic acid.

$$Ca(OH)_2 + HOCH_2CH_2CO_2H \rightarrow Ca(OCOCH_2CH_2OH)_2$$

Calcium lactate is sold for use in medicines and in foods as a source of calcium and as an intermediate in the purification of lactic acid from fermentation processes.

CALCIUM OXIDE

Calcium oxide (CaO, lime, quicklime, unslaked lime) is differentiated from calcium hydroxide [Ca(OH)$_2$, slaked lime, hydrated lime] and limestone (CaCO$_3$, calcite, calcium carbonate, marble chips, chalk) by formula and by behavior. A saturated solution of calcium oxide in water is called *limewater* and a suspension in water is called *milk of lime*.

Lime is manufactured by *calcining* or heating limestone in a kiln.

$$CaCO_3 \rightarrow CaO + CO_2$$

Temperatures used in converting limestone into lime are on the order of 1200 to 1300°C.

In the process (Fig. 1), the limestone (CaCO$_3$) is crushed and screened to a size of approximately 4 to 8 inches. The limestone enters the top of the kiln (Fig. 2) and air enters the bottom and fluidizes the solids for better circulation and reaction. Approximately 98 percent decarbonation is typical.

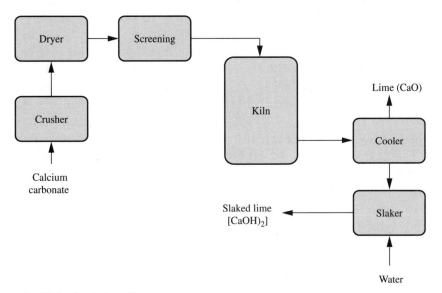

FIGURE 1 Manufacture of lime.

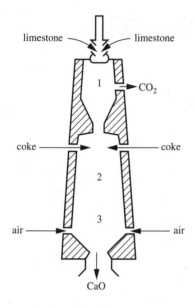

1. *preheating section*

2. *reaction zone*   **FIGURE 2**   A lime kiln.

3. *cooling zone*

When a kiln is used in conjunction with the Solvay process and the manufacture of soda ash, coke can be fired in the kiln along with limestone to give the larger percentages of carbon dioxide needed for efficient soda ash production. If a purer lime product is desired, the fine lime can be taken from area 4 in (Fig. 2). A less pure product is obtained from the bottom kiln section. Another kind of kiln is the rotating, nearly horizontal type. These can be as much as 12 ft in diameter and 450 ft long. Limestone enters one end. It is heated, rotated, and slowly moves at a slight decline to the other end of the kiln, where lime is obtained.

For most applications slaked lime [$Ca(OH)_2$] is sold, since hydration of lime:

$$CaO + H_2O \rightarrow Ca(OH)_2$$

is an exothermic reaction and could ignite paper or wood containers of the unslaked material. Slaked lime is slightly soluble in water to give a weakly basic solution.

Lime is used in the metallurgical industry as a flux in the manufacture of steel. Silicon dioxide is a common impurity in iron ore that cannot be melted unless it combines with another substance first to convert it to a more fluid lava called *slag*. The molten silicate slag is less dense than the molten iron and collects at the top of the reactor, where it can be drawn off.

$$CaO + SiO_2 \rightarrow CaSiO_3$$

Lime is used in water treatment to remove calcium and bicarbonate ions.

$$Ca(OH)_2 + Ca^{2+} + 2HCO^- \rightarrow 2CaCO_3 + 2H_2O$$

Lime is also used in pollution control where lime scrubbers in combustion stacks remove sulfur dioxide present in gases from the combustion of high-sulfur coal.

$$SO_2 + H_2O \rightarrow H_2SO_3$$

$$Ca(OH)_2 + H_2SO_3 \rightarrow CaSO_3 + 2H_2O$$

Lime is also used in the kraft pulping process for the purpose of regenerating caustic soda (sodium hydroxide).

$$Na_2CO_3 + CaO + H_2O \rightarrow CaCO_3 + 2NaOH$$

The caustic soda is then used in the digestion of wood. The lime is regenerated from the limestone by heating in a lime kiln.

A large part of portland cement is lime-based. Sand, alumina, and iron ore are mixed and heated with limestone to 1500°C. Average percentages of the final materials in the cement and their structures are:

20%	$2CaO \cdot SiO_2$	Dicalcium silicate
50	$3CaO \cdot SiO_2$	Tricalcium silicate
10	$3CaO \cdot Al_2O_3$	Tetracalcium aluminate
10	$4CaO \cdot Al_2O_3 \cdot Fe_2O_3$	Tetracalcium aluminoferrite
10	MgO	Magnesium oxide

The percentage of dicalcium silicate determines the final strength of the cement, whereas the amount of tricalcium silicate is related to the early strength (7 to 8 days) required of the cement. Tricalcium aluminate relates to the set in the cement, and tetracalcium aluminoferrite reduces the heat necessary in manufacture.

See **Cement.**

CALCIUM PHOSPHATE

Monobasic calcium phosphate [$Ca(H_2PO_4)_2$] is manufactured by crystallization after evaporation and cooling of a hot solution of lime [CaO or $Ca(OH)_2$] and strong phosphoric acid (H_3PO_4). The crystals are centrifuged, and the highly acidic mother liquor returned for reuse. This acid salt is also made by spray drying a slurry of the reaction product of lime and phosphoric acid.

Dibasic calcium phosphate ($CaHPO_4$) is manufactured from phosphoric acid and lime:

$$CaO + H_3PO_4 \rightarrow Ca_2HPO_4 + H_2O$$

whereas *calcium metaphosphate* is manufactured from phosphate rock, phosphorus pentoxide, and phosphorous acid.

$$CaF_2 \cdot 3Ca_3(PO_4)_2 + 6P_2O_5 + 2HPO_3 \rightarrow 10Ca_3(PO_3)_2 + 2HF$$

In this process the phosphorus pentoxide contacts the lump rock in a vertical shaft.

This calcium metaphosphate may be regarded as a dehydrated triple superphosphate made directly from phosphate rock. The calcium metaphosphate is quite insoluble, and it must hydrolyze to become effective.

$$xCa(PO_3)_2 + xH_2O \rightarrow xCa(H_2PO_4)_2$$

See **Lime.**

CALCIUM SOAPS

Calcium soaps such as calcium stearate, calcium palmitate, and calcium abietate are made by the action of the sodium salts of the acids on a soluble calcium salt such as calcium chloride ($CaCl_2$).

Calcium soaps are insoluble in water but are soluble in hydrocarbons. They tend to form jellylike masses and are used as constituents of greases and as waterproofing agents.

CALCIUM SULFATE

Calcium sulfate occurs in large deposits throughout the world as its hydrate (gypsum; $CaSO_4 \cdot 2H_2O$). When heated at moderate temperatures, gypsum loses part of the water to form a semihydrate.

$$CaSO_4 \cdot 2H_2O \rightarrow CaSO_4 \cdot \tfrac{1}{2}H_2O + 1\tfrac{1}{2}H_2O$$

At higher temperatures, gypsum loses all its water and becomes anhydrous calcium sulfate, or *anhydrite*.

Calcined gypsum (the half-water salt) can be made into wall plaster by the addition of a filler material such as asbestos, wood pulp, or sand. Without additions, it is plaster of Paris and is used for making casts and for plaster.

The usual method of calcination of gypsum consists in grinding the mineral and placing it in large calciners holding up to 20 or more tons. The temperature is raised to about 120 to 150°C, with constant agitation to maintain a uniform temperature. The material in the kettle, *plaster of Paris* or *first-settle plaster*, may be withdrawn at this point, or it may be heated further to 190°C to give a material known as *second-settle plaster*. First-settle plaster is approximately the half-hydrate ($CaSO_4 \cdot \tfrac{1}{2}H_2O$), and the second form is anhydrous ($CaSO_4$). Gypsum may also be calcined in rotary kilns similar to those used for limestone.

CALCIUM SULFIDE

Calcium sulfide is made by reducing calcium sulfate with coke.

$$CaSO_4 + 2C \rightarrow CaS + 2CO_2$$

Calcium sulfide is used as a depilatory in the tanning industry and in cosmetics and, in a finely divided form, it is employed in luminous paints. Calcium polysulfides (CaS_n), such as calcium disulfide (CaS_2) and calcium pentasulfide (CaS_5), are made by heating sulfur and calcium hydroxide.

$$Ca(OH)_2 + S \rightarrow CaS_n + H_2O + SO_x$$

The calcium polysulfides are used as fungicides.

See Calcium Sulfate.

CAPROLACTAM

Caprolactam (melting point: 69.3°C, density: 1.02, flash point 125°C, fire point: 140°C), so named because it is derived from the original name for the C_6 carboxylic acid, caproic acid, is the cyclic amide (lactam) of 6-aminocaproic acid.

Caprolactam is usually manufactured from cyclohexanone, made by the oxidation of cyclohexane or by the hydrogenation/oxidation of phenol (Fig. 1), although the manufacture can be an integrated process with several starting materials (Fig. 2). The cyclohexanol that is also produced with the cyclohexanone can be converted to cyclohexanone by a zinc oxide (ZnO) catalyst at 400°C. The cyclohexanone is converted into the oxime with hydroxylamine, which then undergoes rearrangement to give caprolactam.

Sulfuric acid at 100 to 120°C is often used as the acid catalyst, but phosphoric acid is also used, since after treatment with ammonia the by-product

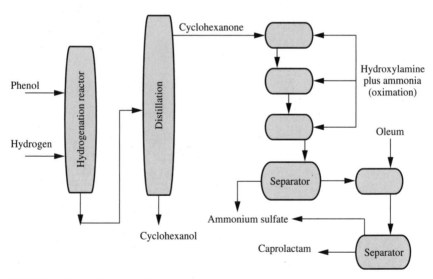

FIGURE 1 Manufacture of caprolactam from phenol.

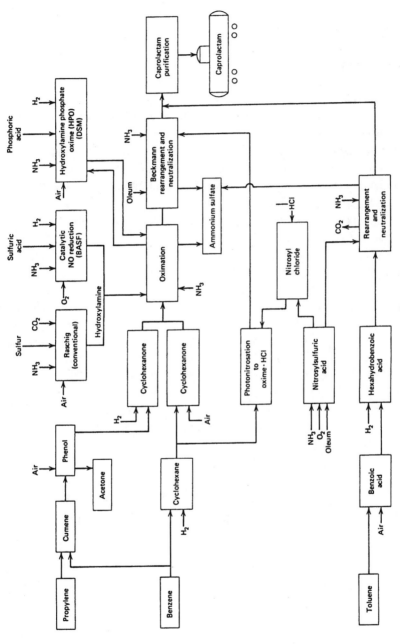

FIGURE 2 Integrated processes for the manufacture of caprolactam from various sources.

becomes ammonium phosphate that can be sold as a fertilizer. The caprolactam can be extracted and vacuum distilled (boiling point: 139°C/12 mmHg) and the overall yield is 90 percent.

Caprolactam is also manufactured from toluene (Fig. 2) by oxidation of toluene (with air) to benzoic acid at 160°C and 10 atm pressure. The benzoic acid is then hydrogenated under pressure (16 atm) and 170°C in a series of continuous stirred tank reactors. The cyclohexane carboxylic acid is blended with oleum and fed to a multistage reactor, where it is converted to caprolactam by reaction with nitrosyl sulfuric acid.

Another process utilizes a photochemical reaction in which cyclohexane is converted into cyclohexanone oxime hydrochloride (Fig. 2), from which cyclohexanone is produced. The yield of cyclohexanone is estimated at about 86 percent by weight. Then, in a rearrangement reaction, the cyclohexanone oxime hydrochloride is converted to e-caprolactam.

Caprolactam is used for the manufacture of nylon 6, especially fibers and plastic resin and film. Nylon 6 is made directly from caprolactam by heating with a catalytic amount of water.

In one process for nylon manufacture, the feedstock is nitration-grade toluene, air, hydrogen, anhydrous ammonia (NH_3), and sulfuric acid (H_2SO_4). The toluene is oxidized to yield a 30% solution of benzoic acid, plus intermediates and by-products. Pure benzoic acid, after fractionation, is hydrogenated with a palladium catalyst in stirred reactors operated at about 170°C under a pressure of 147 psi (1013 kPa). The product, cyclohexanecarboxylic acid, is mixed with sulfuric acid and then reacted with nitrosylsulfuric acid to yield caprolactam.

The nitrosylsulfuric acid is produced by absorbing mixed nitrogen oxides ($NO + NO_2$) in sulfuric acid:

$$NO + NO_2 + H_2SO_4 \rightarrow SO_3 + 2NOHSO_4$$

The resulting acid solution is neutralized with ammonia to yield ammonium sulfate [$(NH_4)_2SO_4$] and a layer of crude caprolactam that is purified according to future use.

CARBON

Carbon exists in several forms: lampblack, carbon black, activated carbon, graphite, and industrial diamonds. The first three types of carbon are examples of *amorphous carbon.*

The term *carbon black* includes furnace black, colloidal black, thermal black, channel black, and acetylene black. Carbon black is mostly derived from petroleum and involves partial combustion or a combination of combustion and thermal cracking of hydrocarbons, and to a lesser extent natural gas, at 1200 to 1400°C.

Lampblack, is soot formed by the incomplete burning of carbonaceous solids or liquids. It is gradually being replaced in most uses by *carbon black* that is the product of incomplete combustion. *Activated carbon* is amorphous carbon that has been treated with steam and heat until it has a very great affinity for adsorbing many materials. *Graphite* is a soft, crystalline modification of carbon that differs greatly in properties from amorphous carbon and from diamond. *Industrial diamonds,* both natural and synthetic, are used for drill points, special tools, glass cutters, wire-drawing dies, diamond saws. and many other applications where this hardest of all substances is essential.

Lampblack, or soot, is an old product and is manufactured by the combustion of petroleum or coal-tar by-products in an oxygen-diminished atmosphere. In the process, the soot is collected in large chambers from which the *raw* lampblack is removed, mixed with tar, molded into bricks, or *pugs,* and calcined up to about 1000°C, after which the product is ground to a fine powder.

The four basic *carbon black* manufacturing processes are either of the partial combustion type (the channel, oil furnace, or gas furnace process) or of the cracking type (the thermal process).

The oil furnace process uses aromatic petroleum oils and residues as feedstock and in the *oil furnace process* (Fig. 1), a highly aromatic feedstock oil (usually a refinery catalytic cracker residue or coal tar-derived material) is converted to the desired grade of carbon black by partial combustion and pyrolysis at 1400 to 1650°C in a refractory (mainly alumina) -lined steel reactor.

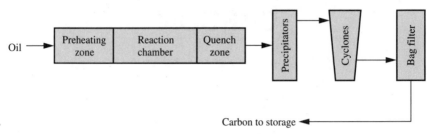

FIGURE 1   Carbon manufacture by the oil furnace process.

The feedstock is atomized and sprayed through a specially designed nozzle into a highly turbulent combustion gas stream formed by preburning natural gas or oil with about a 50 percent excess of air. Insufficient air remains for feedstock combustion, and the oil spray is pyrolyzed in a few microseconds into carbon black with yields (based on carbon) varying from 65 to 70 percent for large carcass-grade types down to about 35 percent for small tread-grade types.

The emissions from the reaction chamber enter the tunnel where it is quenched with water to about 200°C. Tunnel and combustion chamber dimensions vary, but frequently are from 2 to 6 m long and from 0.1 to 1 m in internal diameter. Further downstream, precipitators, cyclones, and to a greater extent bag filters (insulated to remain above 100°C) are used to separate the product black from water vapor and the combustion off-gases. Fiberglass bag filters allow near quantitative yield of the carbon black. Following pulverization and grit removal steps, the loose, fluffy black is normally pelletized for convenient bulk handling and to reduce the tendency to form dust. In the commonly used wet pelletizing process, the black is agitated and mixed with water in a trough containing a rotating shaft with radially projecting pins, which form the pellets. The wet pellets, usually in the 25- to 60-mesh range, are dried in a gas-fired rotating-drum dryer. In the dry pelletizing process, pellets are formed by gentle agitation and rolling of the black in horizontal rotating drums (about 3 m in diameter and 6 to 12 m long) for periods of 12 to 36 hours. An oil pelletizing process is also used, to a minor extent, for specialty applications (e.g., certain ink-grade blacks).

The *channel black process* involves the impingement of natural gas flames on 20- to 25-cm *channel irons* that are slowly reciprocated over scrapers to remove soot deposits. The type of black produced is controlled by burner-tip design, burner-to-channel distance, and air supply (degree of partial combustion).

In the *thermal black process,* natural gas is cracked to carbon black and hydrogen at 1100 to 1650°C in a refractory-lined furnace in a two-cycle (heating and "making," or decomposition) operation. The reaction is

$$CH_4 \rightarrow C + 2H_2$$

Yields of carbon are in the range 30 to 45 percent.

The *gas furnace process,* is similar to the oil furnace process but, like the *thermal black process,* uses natural gas as feedstock.

Activated carbon is manufactured from carbonaceous materials, such as petroleum coke, sawdust, lignite, coal, peat, wood, charcoal, nutshells, and fruit pits. Activation is a physical change wherein the surface of the carbon is increased by the removal of hydrocarbons by any one of several methods. The most widely used methods involve treatment of the carbonaceous material with oxidizing gases such as air, steam, or carbon dioxide, and the carbonization of the raw material in the presence of chemical agents such as zinc chloride or phosphoric acid.

After activated carbon has become saturated with a vapor or an adsorbed color, either the vapor can be steamed out, condensed, and recovered (Fig. 2), or the coloration can be destroyed and the carbon made ready for reuse. The oldest example of this process uses the decolorizing carbon long known as *bone char,* or *bone black.* This consists of about 10% carbon deposited on a skeleton of calcium phosphate $[Ca_3(PO_4)_2]$ and is made by the carbonization of fat-free bones in closed retorts at 750 to 950°C.

Graphite, a naturally occurring form of carbon, has been known for many centuries and occurs throughout the world in deposits of widely

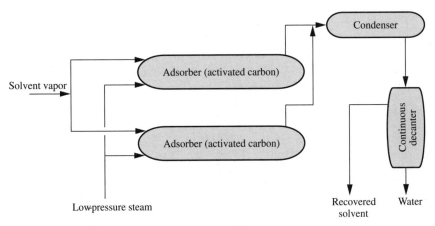

FIGURE 2 Solvent recovery process.

varying purity and crystalline size and perfection. For import purposes, natural graphite is classified as *crystalline* and *amorphous*. The latter is not truly amorphous but has an imperfect lamellar microcrystalline structure.

Graphite is made electrically from retort or petroleum coke (Fig. 3). Temperatures on the order of 2700°C are necessary.

$$C \text{ (amorphous)} \rightarrow C \text{ (graphite)}$$

In the process, the carbon feedstock is calcined (1250°C) to volatilize any impurities, after which the calcined products are ground, screened, weighed, mixed with binder, formed by molding or extrusion into green electrodes, and arranged in the furnace (Fig. 4).

In the manufacture of industrial diamonds, the process is a batch process and requires pressures and temperatures in the region of thermodynamic stability for diamond and a molten catalyst-solvent metal consisting of a group VIII metal or alloy. Special ultra high-pressure apparatus is used, the moving members of which are forced together by large hydraulic presses. Different types and sizes of diamond particles, or crystals, require different conditions of pressure, temperature, catalyst-solvent, and reaction time. The crude diamonds are cleaned and graded by size and shape.

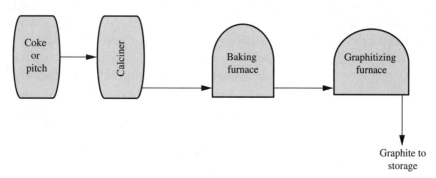

FIGURE 3   Graphite manufacture.

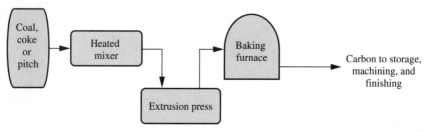

FIGURE 4   Manufacture of carbon electrodes.

Carbon black has uses similar to some inorganic compounds. For example, it is the most widely used black pigment. It is also an important reinforcing agent for various elastomers. It is used in tires and other elastomers and is a leading pigment in inks and paints.

Carbon black is a black pigment, and lampblack, which has a larger particle size, is used for tinting to produce shades of gray. Carbon black is very opaque and has excellent durability, resistance to all types of chemicals, and light fastness.

CARBON BLACK

*See **Carbon.***

CARBON DIOXIDE

Carbon dioxide (CO_2, melting point: $-56.6°C$ at 76 psi – 527 kPa, density: 1.9769 gL at $0°C$) is a colorless, odorless gas. Solid carbon dioxide sublimes at $-79°C$ (critical pressure: 1073 psi, 7397 kPa, critical temperature: $31°C$). High concentrations of the gas do cause stupefaction and suffocation because of the displacement of ample oxygen for breathing. Carbon dioxide is soluble in water (approximately 1 volume carbon dioxide in 1 volume water at $15°C$), soluble in alcohol, and is rapidly absorbed by most alkaline solutions.

Carbon dioxide is made by steam-reforming hydrocarbons (Fig. 1), and much of the time natural gas is the feedstock.

$$CH_4 + 2H_2O \rightarrow 4H_2 + CO_2$$

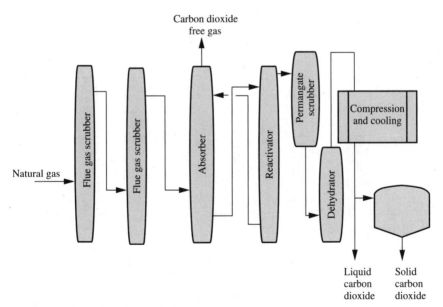

FIGURE 1 Manufacture of carbon dioxide.

A small amount of carbon dioxide is still made from fermentation of grain; ethyl alcohol is the main product (Fig. 2).

$$C_6H_{12}O_6 \rightarrow 2C_2H_5OH + 2CO_2$$

Other small amounts are obtained from coke burning, the calcination of lime, in the manufacture of sodium phosphates from soda ash and phosphoric acid, by recovery from synthesis gas in ammonia production, recovery as a by-product in the production of substitute natural gas, and recovery from natural wells.

Although carbon dioxide must be generated on site for some processes, there is a trend toward carbon dioxide recovery where it is a major reaction by-product and, in the past, has been vented to the atmosphere. An absorption system, such as the use of ethanolamines (*q.v.*) or hot carbonate or bicarbonate solutions, is used for concentrating the carbon dioxide to over 99% purity.

$$Na_2CO_3 + CO_2 + H_2O \rightarrow 2NaHCO_3$$

$$2NaHCO_3 \rightarrow Na_2CO_3 + CO_2 + H_2O$$

In all cases, the almost pure carbon dioxide must be given various chemical treatments for the removal of minor impurities that contaminate it.

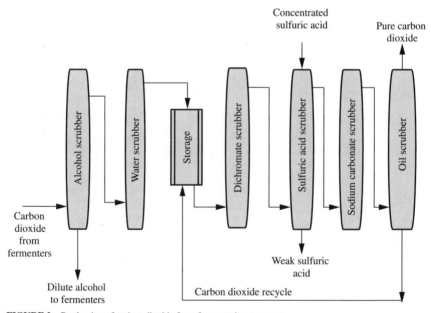

FIGURE 2 Production of carbon dioxide from fermentation processes.

Carbon dioxide, in the solid form, is commonly referred to as *dry ice*. At that temperature it sublimes and changes directly from a solid to a vapor. Because of this property, as well as its noncombustible nature, it is a common refrigerant and inert blanket. Carbon dioxide is also used for enhanced oil recovery.

Other uses include chemical manufacture (especially soda ash), fire extinguishers, and pH control of wastewater. Carbon dioxide is also used as a fumigant for stored grain and has replaced ethylene dibromide in this application.

CARBON MONOXIDE

Carbon monoxide (melting point: $-199°C$, boiling point: $-191.5°C$) is one of the chief constituents of synthesis gas (carbon monoxide plus hydrogen). It is obtained in pure form through cryogenic procedures, with hydrogen as a coproduct.

Carbon monoxide is an important raw material in the production of methanol and other alcohols and of hydrocarbons, and it is a powerful poison. It is also used for making diisocyanate and ethyl acrylate.

$$2CO + 2Cl_2 \rightarrow 2COCl_2$$

$$2COCl_2 + \text{toluene–2,4–diamine} \rightarrow CH_3C_6H_3(NCO)_2 + 4HCl$$
toluene diisocyanate

$$CO + HC{\equiv}CH + C_2H_5OH \rightarrow CH_2{=}CHCOOC_2H_5$$
ethyl acrylate

See also **Synthesis Gas.**

CARBON TETRACHLORIDE

Carbon tetrachloride (melting point: $-23°C$, boiling point: $76.7°C$, density: 1.5947, critical temperature $283.2°C$, critical pressure 9714 psi – 67 MPa, solubility 0.08 g in 100 g water) is a heavy, colorless, nonflammable, noncombustible liquid. Dry carbon tetrachloride is noncorrosive to common metals except aluminum. When wet, carbon tetrachloride hydrolyzes and is corrosive to iron, copper, nickel, and alloys containing those elements.

Carbon tetrachloride is manufactured by the reaction of carbon disulfide and chlorine, with sulfur monochloride as an important intermediate.

$$CS_2 + 3Cl_2 \rightarrow S_2Cl_2 + CCl_4$$

$$CS_2 + 2S_2Cl_2 \rightarrow 6S + CCl_4$$

The reaction must be carried out in a lead-lined reactor in a solution of carbon tetrachloride at $30°C$ in the presence of iron filings as catalyst. The elemental sulfur can be reconverted to carbon disulfide by reaction with coke.

$$6S_4 + 3C \rightarrow 3CS_2$$

Chlorination of methane and higher aliphatic hydrocarbons is also used to manufacture carbon tetrachloride.

$$CH_4 + 4Cl_2 \rightarrow CCl_4 + 4HCI$$

The chlorination of methane is also used to manufacture partially chlorinated methane derivatives.

$$4CH_4 + 10Cl_2 \rightarrow CH_3Cl + CH_2Cl_2 + CHCl_3 + CCl_4 + 10HCl$$

The reaction is carried out in the liquid phase at about $35°C$, and ultraviolet light is used as a catalyst. The same reaction can be carried out at $475°C$ without catalyst. The unreacted methane and partially chlorinated products are recycled to control or adjust the yield of carbon tetrachloride.

CELLULOSE

Cellulose is the primary substance of which the walls of vegetable cells are constructed and is largely composed of glucose residues. It may be obtained from wood or derived in very high purity from cotton fibers, which are about 92% pure cellulose.

The important fiber rayon is *regenerated* cellulose from wood pulp that is in a form more easily spun into fibers. Cellophane film is regenerated cellulose made into film. One method of regeneration is formation of xanthate groups from selected hydroxyl groups of cellulose, followed by hydrolysis back to hydroxy groups.

Cellulose acetate and triacetate may be used as plastics or spun into fibers for textiles. They are made by the reaction of cellulose with acetic anhydride.

*See **Cellulose Acetate, Cellulose Nitrate.***

CELLULOSE ACETATE

Fully acetylated cellulose is partly hydrolyzed to give an acetone-soluble product, which is usually between the di- and triester. The esters are mixed with plasticizers, dyes, and pigments and processed in various ways, depending upon the form of plastic desired.

Cellulose acetate is moldable, in contrast to cellulose nitrate, which is not. It, however, has poor resistance to moisture. To overcome this disadvantage, treating the cotton linters with a mixture of butyric acid, acetic acid, and acetic anhydride produces a mixed acetate-butyrate. The plastics made from this material have better moisture resistance and better dimensional stability than cellulose acetate. They also have excellent weathering resistance, high impact strength, moldability, and high dielectric strength.

CELLULOSE NITRATE

Fully nitrated cellulose (nitrocellulose) is unsuited for a plastic base because of its extremely flammable character. Hence, a partly nitrated product is made. Products containing 11% nitrogen are used for plastics, 12% nitrogen for lacquers, and the fully nitrated, or 13% nitrogen, material is used for explosives.

In the process (Fig. 1), the cellulose nitrate is placed in large kneading mixers with solvents and plasticizers and thoroughly mixed. The standard plasticizer is camphor, and the compounded mixture is strained under hydraulic pressure and mixed on rolls with coloring agents. The material is pressed into blocks.

Finally, the plastic is made into sheets, strips, rods, or tubes, seasoned to remove the residual solvent, and polished by pressing under low heat. This plastic possesses excellent workability, water resistance, and toughness. Its chief disadvantage is the ease with which it burns. It also discolors and becomes brittle on aging. The trade name is Celluloid.

*See **Cellulose, Cellulose Acetate.***

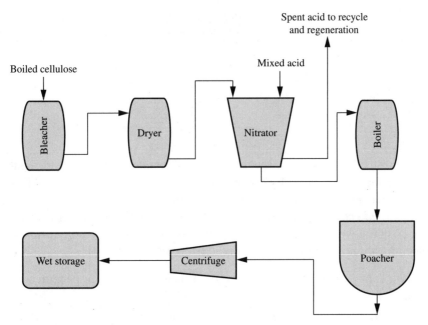

FIGURE 1 Manufacture of cellulose nitrate.

CEMENT

The terms *cement* and *concrete* are not synonymous. Concrete is artificial stone made from a carefully controlled mixture of cement, water, and fine and coarse aggregate (usually sand and coarse rock).

Portland cement is the product obtained by pulverizing clinker consisting essentially of calcium silicate, usually containing one or more forms of calcium sulfate, and there are five types of portland cement:

Type I. Regular portland cements are the usual products for general construction.

Type II. Moderate-heat-of-hardening and *sulfate-resisting* portland cements are for use where moderate heat of hydration is required or for general concrete construction exposed to moderate sulfate action.

Type III. High-early-strength (HES) cements are made from raw materials with a lime-to-silica ratio higher than that of Type I cement and are ground finer than Type I cements. They contain a higher proportion of calcium silicate than regular portland cements.

Type IV. Low-heat portland cements contain a lower percentage of calcium silicate and calcium aluminate, thus lowering the heat evolution.

Type V. Sulfate-resisting portland cements are those that, by their composition or processing, resist sulfates better than the other four types. Type V is used when high sulfate resistance is required.

Two types of materials are necessary for the production of portland cement: one rich in calcium (calcareous), such as limestone or chalk, and one rich in silica (argillaceous) such as clay. These raw materials are finely ground, mixed, and heated (burned) in a rotary kiln to form cement clinker.

Cement clinker is manufactured by both the *dry process* and by the *wet process* (Fig. 1). In both processes *closed-circuit grinding* is preferred to *open-circuit grinding* in preparing the raw materials because, in the former, the fines are passed on and the coarse material returned, whereas in the latter, the raw material is ground continuously until its mean fineness has reached the desired value. The *wet process,* though the original one, is

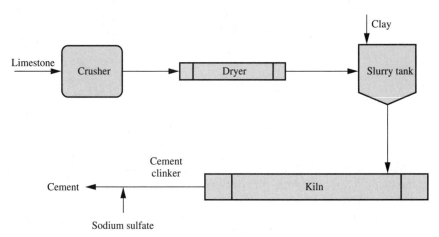

FIGURE 1 Cement manufacture.

being displaced by the *dry process,* especially for new plants, because of the savings in heat and accurate control and mixing of the raw mixture it affords.

The *dry process* is applicable to natural cement rock and to mixtures of limestone and clay, shale, or slate. In this process the materials may be roughly crushed, passed through gyratory and hammer mills, dried, sized, and more finely ground in tube mills, followed by air separators, after which thorough mixing and blending by air takes place. This dry, powdered material is fed directly to rotary kilns that are slightly inclined, so that materials fed in at the upper end travel slowly to the lower firing end, taking from 1 to 3 hours.

In the *wet process,* the solid material, after dry crushing, is reduced to a fine state of division in wet tube or ball mills and passes as a slurry through bowl classifiers or screens. The slurry is pumped to correcting tanks, where rotating arms make the mixture homogeneous and allow the final adjustment in composition to be made.

The final product consists of hard, granular masses from 3 to 20 mm in size, called *clinker.* The clinker is discharged from the rotating kiln into the air-quenching coolers that reduce the temperature to approximately 100 to 200°C while simultaneously preheating the combustion air. Pulverizing, followed by fine grinding in the tube ball mills and automatic packaging, completes the process. During the fine grinding, setting retarders, such as gypsum, plaster, or calcium lignosulfonate, and air-entraining, dispersing, and waterproofing agents are added. The clinker is ground dry by various hookups.

CEPHALOSPORINS

Cephalosporins contain the β-lactams structure, like the penicillins, but instead of a five-membered thiozolidine ring, cephalosporins contain a six-membered dihydrothiazine ring.

Cephalosporin C, which itself is not antibacterial, is obtained from a species of fungus. Chemical modification of this structure to 7-aminocephalosporanic acid by removal of the R-C=O group allows the preparation of the active cephalosporins such as cephalexin and cephaloglycin.

The drugs are orally active because they have an a-amino-containing R group that is stable to the gastric acid in the stomach. Other cephalosporins are easily made by acylation of the 7-amino group with different acids or nucleophilic substitution or reduction of the 3-acetoxy group.

CHLORAL

Chloroacetaldehydes are produced by the reaction of acetaldehyde with chlorine by replacement of the hydrogen atoms of the methyl group. Chlorine reacts with acetaldehyde at room temperature to give monochloroacetaldehyde. Increasing the temperature to 70 to 80°C gives dichloroacetaldehyde, and at a temperature of 80 to 90°C trichloroacetadehyde, (chloral) is formed.

$$CH_3CH=O + Cl_2 \rightarrow CH_2ClCH=O + HCl$$

$$CH_2ClCH=O + Cl_2 \rightarrow CHCl_2CH=O + HCl$$

$$CHCl_2CH=O + Cl_2 \rightarrow CCl_3CH=O + HCl$$

Bromal is formed by an analogous series of reactions.

CHLORINATED SOLVENTS

Originally, successive chlorination and dehydrochlorination of acetylene was the route to trichloroethylene (Cl_2C=CHCl):

$$HC{\equiv}CH + 2Cl_2 \rightarrow CHCl_2CHCl_2$$

$$CHCl_2CHCl_2 \rightarrow CHCl{=}CCl_2$$

$$CHCl{=}CCl_2 + Cl_2 \rightarrow CH\,Cl_2{=}CCl_3$$

$$CH\,Cl_2{=}CCl_3 \rightarrow CCl_2{=}CCl_2$$

This route has largely been superceded by a route involving chlorination and dehydrochlorination of ethylene or vinyl chloride and, more recently, by a route involving oxychlorination of two-carbon (C_2) raw materials.

CHLORINE

Chlorine (Cl_2, melting point: $-101\,°C$, boiling point: $-34.6\,°C$, density (gas): 3.209 g/L at $0\,°C$) is a pale greenish-yellow gas of marked odor, irritating to the eyes and throat, and poisonous.

Chlorine is produced almost entirely by the electrolysis of aqueous solutions of alkali metal chlorides (Fig. 1), or from fused chlorides. Brine electrolysis produces chlorine at the anode and hydrogen along with the alkali hydroxide at the cathode. At present, three types dominate the industry: the diaphragm cell, the membrane cell, and the mercury cell, and there are many variations of each type.

In production of chlorine by the diaphragm cell process (Fig. 1), salt is dissolved in water and stored as a saturated solution. Chemicals are added to adjust the pH and to precipitate impurities from both the water and the

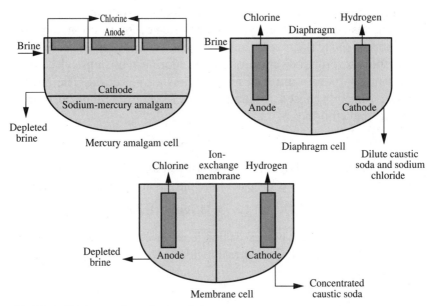

FIGURE 1 Chlorine manufacture by electrolysis.

salt. Recycled salt solution is added. The precipitated impurities are removed by settling and by filtration. The purified, saturated brine is then fed to the cell, which typically is a rectangular box. It uses vertical anodes (ruthenium dioxide with perhaps other rare metal oxides deposited on an expanded titanium support). The cathode is perforated metal that supports the asbestos diaphragm that has been vacuum deposited in a separate operation. The diaphragm serves to separate the anolyte (the feed brine) from the catholyte (brine containing caustic soda). Chlorine is evolved at the anode. It is collected under vacuum, washed with water to cool it, dried with concentrated sulfuric acid, and further scrubbed, if necessary. It is then compressed and sent to process as a gas or liquefied and sent to storage.

The membrane cell (Fig. 1) uses a cation exchange membrane in place of an asbestos diaphragm. It permits the passage of sodium ions into the catholyte but effectively excludes chloride ions. Thus the concept permits the production of high-purity, high-concentration sodium hydroxide directly.

In the mercury cell process (Fig. 1), chlorine is liberated from a brine solution at the anode. Collection and processing of the chlorine is similar to the techniques employed when diaphragm cells are used. However, the cathode is a flowing bed of mercury. When sodium is released by electrolysis it is immediately amalgamated with the mercury. The mercury amalgam is then decomposed in a separate cell to form sodium hydroxide and the mercury is returned for reuse.

Other processes for making chlorine include sodium manufacture, caustic potash manufacture, hydrogen chloride decomposition, the nitrosyl chloride (NOCl) process, and a process where salt is treated with nitric acid to form sodium nitrate and chlorine with nitrosyl chloride (containing 4 to 10% nitrogen tetroxide) as a by-product. The nitrosyl chloride vapor is placed in contact with oxygen to produce nitrogen tetroxide and chlorine:

$$2NOCl + O_2 \rightarrow N_2O_4 + Cl_2$$

After liquefying and distilling the chlorine out, the nitrogen tetroxide is absorbed in water to make nitric acid and nitrous acids, which are recycled:

$$N_2O_4 + H_2O \rightarrow HNO_3 + HNO_2$$

The advantage of this process is that it produces chlorine but no caustic soda. The demand for sodium nitrate regulates the amount of chlorine that can be made in this way.

$$3NaCl + 4HNO_3 \rightarrow 3NaNO_3 + Cl_2 + NOCl + 2H_2O$$

$$2NOCl + 3HNO_2 + 3O_2 + H_2O \rightarrow 5HNO_3 + Cl_2$$

Chlorine is principally used to produce organic compounds. But, in many cases chlorine is used as a route to a final product that contains no chlorine. For instance propylene oxide has traditionally been manufactured by the chlorohydrin process. Modern technology permits abandoning this route in favor of direct oxidation, thus eliminating a need for chlorine.

Chlorine is also used in bleaching and for treating municipal and industrial water supplies, and this use will probably continue. However, some concern has been felt that traces of organic compounds in all water supplies react with the chlorine to form chlorinated organics which are suspected of being carcinogenic. Further, the usefulness of chlorination of municipal wastes has been questioned in some quarters in the light of the fact that such treatment adds chlorinated organics to the waterways.

CHLORINE DIOXIDE

See **Sodium Chlorite.**

CHLOROACETALDEHYDE

See **Chloral.**

CHLOROFLUOROCARBONS

Chlorofluorocarbons (CFCs) are manufactured by reacting hydrogen fluoride and carbon tetrachloride in the presence of a partially fluorinated antimony pentachloride catalyst in a continuous, liquid-phase process.

$$2CCl_4 + 3HF + SbCl_2F_2 \rightarrow CCl_2F_2 + CCl_3F + 3HCl$$

The products dichlorodifluoromethane and trichlorofluoromethane are also known as CFC-12 and CFC-11, respectively.

There are common abbreviations and a numbering system for chlorofluorocarbons and related compounds. The original nomenclature developed in the 1930s is still employed and uses three digits. When the first digit is 0, it is dropped. The first digit is the number of carbon atoms minus 1, the second digit is the number of hydrogen atoms plus 1, and the third digit is the number of fluorine atoms. All other atoms filling the four valences of each carbon are chlorine atoms. Important nonhydrogen-containing chlorofluorocarbons (Freons®) are:

CCl_2F_2 (CFC-12) CCl_3F (CFC-11) CCl_2FCClF_2 (CFC-113)

Halons, a closely related type of chemical that also contains bromine, are used as fire retardants. Numbering here is more straightforward: first digit, number of carbons; second digit, number of fluorine atoms; third digit, number of chlorines; and fourth digit, number of bromines.

Common Halons are: Halon 1211 (CF_2BrCl), 1301 (CF_3Br), and Halon 2402 (C_2F_4Br).

Compared to the same size hydrocarbons, fluorocarbons have higher volatility and lower boiling points, unusual for halides. They are less reactive, more compressible, and more thermally stable than hydrocarbons. They also have low flammability and odor.

Common uses for the fluorocarbons are as refrigerants, foam-blowing agents, solvents, and fluoropolymers. Recent environmental legislation has restricted or banned the use of chlorofluorocarbons.

CHLOROFORM

Chloroform (boiling point: 61.7°C, melting point: –63.5°C, density: 1.4832) is produced by the chlorination of methylene chloride, which in turn is made by the chlorination of methyl chloride and methane.

$$CH_4 + Cl_2 \rightarrow CH_3Cl + HCl$$

$$CH_3Cl + Cl_2 \rightarrow CHCl_2 + HCl$$

$$CH_2Cl_2 + Cl_2 \rightarrow CHCl_3 + HCl$$

Chloroform is also produced from acetone and calcium hypochlorite—the reaction is rapid and the yield is high.

$$2CH_3COCH_3 + 3Ca(OCl)_2 \rightarrow 2CHCl_3 + Ca(CH_3COO)_2 + 2Ca(OH)_2$$

Pure chloroform decomposes readily on storing, particularly if exposed to moisture and sunlight, to yield phosgene and other compounds. A small amount (0.5 to 1%) of ethyl alcohol is added to retard this decomposition.

The main use of chloroform is in the manufacture of chlorofluorocarbon refrigerants and polymers.

CHLOROPRENE

Chloroprene (boiling point: 59.4°C, density: 0.9583) is, chemically, a chlorovinyl ester of hydrochloric acid and can be manufactured by polymerizing acetylene to vinyl acetylene using a weak solution containing ammonium chloride (NH_4Cl), cuprous chloride (Cu_2Cl_2), and potassium chloride (KCl) as catalyst. The off-gas from the reactor has its water condensed out and is then fractionated. Aqueous hydrochloric acid at 35 to 45°C is then reacted with the vinyl acetylene in the presence of cupric chloride to give chloroprene (2-chloro-1,3-butadiene).

$$2HC{\equiv}CH \rightarrow CH_2{=}CHC{\equiv}CH$$

$$CH_2{=}CHC{\equiv}CH + HCl \rightarrow CH_2{=}CClC{=}CH_2$$

The contact time is about 15 seconds with a 20 percent conversion per pass. An overall yield of approximately 65 percent can be achieved.

Chloroprene is also made by chlorination of butadiene at 300°C followed by dehydrochlorination, using sodium hydroxide at 100°C. Addition of the chlorine to the butadiene occurs at either 1,2 or 1,4 because the intermediate allyl carbocation is delocalized.

$$CH_2{=}CHCH{=}CH_2 + Cl_2 \rightarrow CH_2ClCH{=}CHCH_2Cl$$

$$CH_2ClCH{=}CHCH_2Cl + Cl_2 \rightarrow CH_2{=}CHC(Cl){=}CH_2$$

The 1,4-dichloro isomer can be isomerized to the 1,2-dichloro isomer by heating with cuprous chloride.

$$CH_2ClCH{=}CHCH_2Cl + \rightarrow CH_3CH_2CCl{=}CHCl$$

CHROMIC OXIDE

Chromium oxide (chromic oxide, CrO_3) is one of the oldest known green pigments. It is manufactured by calcining either sodium or potassium dichromate with sulfur in a reverberatory furnace:

$$Na_2Cr_2O_7 + S \rightarrow Cr_2O_3 + Na_2SO_4$$

Hydrated chromic oxide [$Cr_2O(OH)_4$, $Cr_2O_3 \cdot 2H_2O$] is known as Guignet's green (emerald green) and possesses a much more brilliant green color than the oxide and has good permanency.

Chromic oxide is manufactured by roasting a mixture of sodium dichromate and boric acid at a dull red heat for several hours.

CIMETIDINE

Cimetidine, a highly substituted guanidine, is sold as Tagamet, and is widely used as an antiulcer medication. It acts by blocking the histamine molecules in the stomach from signaling the stomach to secrete acid.

Cimetidine is manufactured by the action of a substituted guanidine on an amino-thio compound in the presence of methyl cyanide.

$$CH_3NHCSCH_3 + NH_2CH_2CH_2SCH_2Z \rightarrow CH_3NHCNHCH_2CH_2SCH_2Z$$

CINNAMIC ALDEHYDE

Cinnamic aldehyde (boiling point: 253°C with decomposition, melting point: −7.5°C, density: 1.0497) has a cinnamon odor, hence the name. As it oxidizes in air to cinnamic acid, it should be protected from oxidation.

Although this aldehyde is obtained from Chinese cassia oils, it is synthesized by action of alkali upon a mixture of benzaldehyde and acetaldehyde (Fig. 1).

$$C_6H_5CH=O + CH_3CH=O \rightarrow C_6H_5CH=CHCH=O$$

This and most other products for fragrances must be purified by, for example, vacuum fractionation.

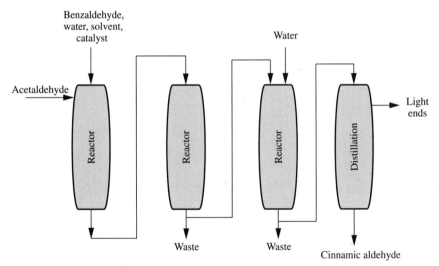

FIGURE 1   Manufacture of cinnamic aldehyde.

CITRIC ACID

Citric acid (melting point: 153°C, density: 1.665) is one of our most versatile organic acids and is used as an acidulant in carbonated beverages, jams, jellies, and other foodstuffs. Another large outlet is in the medicinal field, including the manufacture of citrates and effervescent salts. Industrially citric acid is used as an ion-sequestering agent buffer and in the form of acetyl tributyl citrate, as a plasticizer for vinyl resins.

Except for small amounts produced from citrus-fruit wastes, citric acid is manufactured by aerobic fermentation of crude sugar or corn sugar by a special strain of *Aspergillus niger*,

$$\underset{\text{sucrose}}{C_{12}H_{22}O_{11}} + H_2O + 3O_2 \rightarrow \underset{\text{citric acid}}{2C_6H_8O_7} + 4H_2O$$

$$\underset{\text{dextrose}}{2C_6H_{12}O_6} + 3O_2 \rightarrow \underset{\text{citric acid}}{C_6H_8O_7} + 4H_2O$$

The submerged process for the manufacture of citric acid (Fig. 1) involves coordinated sequences of biochemical conversions with the aid of *A. niger* and various unit operations and chemical conversions.

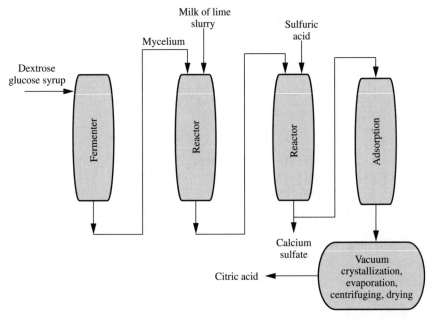

FIGURE 1   Manufacture of citric acid.

COAL CHEMICALS

Coal is an organic rock that can be converted by heat treatment into a variety of products (Fig. 1).

When coal is thermally pyrolyzed or distilled by heating without contact with air, it is converted into a variety of solid, liquid, and gaseous products. The nature and amount of each product depends upon the temperature used in the pyrolysis and the variety of the coal used. In general practice, coke-oven temperatures are maintained above 900°C but may range anywhere from 500 to 1000°C. The principal product by weight is coke.

If temperatures on the order of 450 to 700°C are employed, the process is termed *low-temperature carbonization*; with temperatures above 900°C, it is designated *high-temperature carbonization*. In the low-temperature carbonization process, the quantity of gaseous products is small and that of the liquid products is relatively large, whereas in high-temperature carbonization the yield of gaseous products is larger than the yield of liquid products, the production of tar being relatively low.

For the same coal, low-temperature liquids contain more tar acids and tar bases than high-temperature liquids. With high-temperature carbonization, the liquid products are water, tar, and crude light oil. The gaseous products are hydrogen, methane, ethylene, carbon monoxide, carbon dioxide, hydrogen sulfide, ammonia, and nitrogen. The products other than coke are collectively known as coal chemicals, or *by-products*.

The gas from the destructive distillation of the coal, together with entrained liquid particles, passes upward through a cast-iron gooseneck into a horizontal steel pipe, which is connected to a series of ovens. As the gas leaves the ovens, it is sprayed with weak ammonia water to condense some of the tar from the ammonia from the gas into the liquid. The liquids move through the main along with the gases until a settling tank is reached, where separation is effected according to density. Some of the ammonia liquor is pumped back into the pipes to help condensation; the rest goes to the ammonia still (Fig. 2), which releases the ammonia for subsequent chemical combination in the saturator. All the tar flows to storage tanks for tar distillers or for fuel.

Starting material	Initial products	Secondary products	Tertiary products
Coal	Gas	Fuel gas	
		Sulfur	
	Light oil	Naphtha	Cyclopentadiene
			Indene
		Benzol	Benzene
		Toluols	Toluene
		Xylols	o-xylene
			m-xylene
			p-xylene
	Coal tar	Pitch	
		Tar	Fluorene
			Diphenylene oxide
			Acenaphthene
			Methyl-naphthalenes
		Creosote	
		Tar acids	Phenol
			o-cresol
			m-cresol
			p-cresol
			3, 5-xylenol
		Tar bases	Pyridine
			Alpha-picoline
			Beta-picoline
			Gamma-picoline
			2, 6-lutidine
			Quinoline
		Naphthalene	
		Crude anthracene	Anthracene
			Phenanthrene
			Carbazole
	Coke		Carbon

FIGURE 1 Selected chemicals from coal.

The tar separated from the collecting main and the tar extractors or electrostatic precipitators is settled from ammonia liquor, together with light oil. The tar is filtered and pumped through the reflux vapor-tar heat exchanger and economizer, pitch-tar heat exchanger into the top of the lower third of the distillation column and out at the bottom, to the circulating pumps and into the pipe still where the crude tar joins 4 to 5 volumes of the circulating pitch, and on finally to near the top of the distillation column.

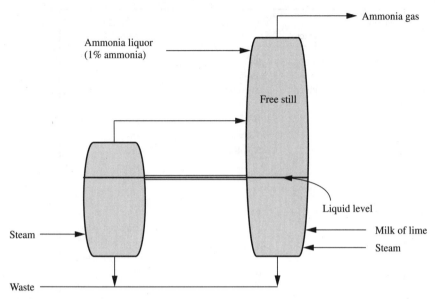

FIGURE 2   An ammonia still.

The vapors, steam distilled and superheated, pass overhead from the top of the tray-type distillation column, enter the bottom of the bubble-cap fractionating column, and are separated into four fractions and a residue, leaving the residue at the bottom of this fractionating column. The pitch cascades from the top of the distillation column down through the super-heated steaming section to establish the desired pitch hardness and to strip this pitch of the higher-boiling volatile oils. It is then withdrawn from above the middle of the distillation column and conducted through the pitch-tar heat exchanger to storage. The products are:

Product	Boiling point, °C
Light oil	To 170
Carbolic oil	170–205
Naphthalene oil	205–240
Creosote or wash oil	240–280
Residue or anthracene	270–340
Residuum or pitch	325–400

The gaseous mixture leaving the oven is made up of permanent gases that form the final purified coke-oven coal gas for fuel, along with condensable water vapor, tar, and light oils, and solid particles of coal dust, heavy hydrocarbons, and complex carbon compounds. Significant products recoverable from the vapors include: benzene, toluene, xylenes, creosote oils, cresols, cresylic acid, naphthalene, phenols, xylols, pyridine, quinoline, and medium and hard pitches usable for electrode binders, road tar, or roofing pitch.

The gas is passed into the primary condenser and cooler at a temperature of about 75°C. Here the gases are cooled by water to 30°C. The gas is conducted to an exhauster, which serves to compress it. During the compression the temperature of the gas rises as high as 50°C. The gas is passed to a final tar extractor or an electrostatic precipitator. On leaving the tar extractor or precipitator, the gas carries three-fourths of the ammonia and 95 percent of the light oil originally present.

The gas is led to a saturator containing a solution of 5 to 10% sulfuric acid where the ammonia is absorbed, and crystalline ammonium sulfate is formed. The gas is fed into the saturator through a serrated distributor underneath the surface of the acid liquid. The acid concentration is maintained by the addition of sulfuric acid, and the temperature is kept at 60°C by the reheater and the heat of neutralization. The crystallized ammonium sulfate is removed from the bottom of the saturator by a compressed-air ejector, or a centrifugal pump, and drained on a table, from which the mother liquor is run back into the saturator.

The gas leaving the saturator is at about 60°C; it is taken to final coolers or condensers, where it is scrubbed with water until its temperature is 25°C. During this cooling, some naphthalene separates and is carried along with the wastewater and recovered. The gas is passed into a light oil or benzol scrubber, through which the absorbent medium, a heavy fraction of petroleum known as straw oil, or sometimes tar oil, is circulated at 25°C. The straw oil is sprayed into the top of the absorption tower while the gas flows upward through the tower. The straw oil is allowed to absorb about 2 to 3 percent of its weight of light oil, with a removal efficiency of about 95 percent of the light oil vapor in the gas.

The rich straw oil, after being warmed in heat exchangers by vapors from the light-oil still and then by light oil flowing out of the still, is passed to the stripping column where the straw oil, flowing downward, is brought into direct contact with live steam. The vapors of the light oil and steam pass off and upward from the still through the heat exchanger previously mentioned and into a condenser and water separator. The lean, or stripped, straw oil is returned through the heat exchanger to the scrubbers. The gas, after having been stripped of its ammonia and light oil, has the sulfur removed in purifying boxes, which contain iron oxide on wood shavings, or by a solution of ethanolamine (Girbotol) in scrubbing towers (the best present-day practice).

An alternative procedure uses ammonium phosphate ($NH_4H_2PO_4$) to absorb the ammonia to form more alkaline phosphates [$(NH_4)_2HPO_4$ and $(NH_4)_3PO_4$] that are returned to the original form by steaming, thus releasing the ammonia (the Phosam process).

Coal tar is a mixture of many chemical compounds, mostly aromatic, which vary widely in composition and can be separated into various fractions by distillation.

1. Light oils usually comprise the cut up to 200°C. They are first crudely fractionated and agitated at a low temperature with concentrated sulfuric acid, neutralized with caustic soda, and redistilled, furnishing benzene, toluene, and homologs.

2. Middle oils, or creosote oils, generally are the fraction from 200 to 250°C, which contain naphthalene, phenol, and cresols. The naphthalene settles out on cooling, is separated by centrifuging, and is purified by sublimation. After the naphthalene is removed, phenol and other tar acids (phenol and its homologs) are obtained by extraction with 10% caustic soda solution and neutralization, or *springing* by carbon dioxide. These are fractionally distilled.

3. Heavy oil may represent the fraction from 250 to 300°C, or it may be split between the middle oil and the anthracene oil.

4. Anthracene oil is usually the fraction from 300 to 350°C. It is washed with various solvents to remove phenanthrene and carbazole; the remaining solid is anthracene.

A substantial fraction of the coal tar produced continues to be used as fuel as well as for road asphalt and roofing asphalt. For these purposes the tar is distilled up to the point where thermal decomposition starts. This *base tar* is then diluted with creosote oil to ensure satisfactory rapid drying. Similar tars are used to impregnate felt and paper for waterproofing materials.

Largely because of the present competition from aromatic chemicals produced from petroleum, interest in aromatics from coal is not as great as it used to be. At one time coal tar was the sole source of pyridine; however, synthetic processes using aldehyde arid ammonia are now supplying the increased demand. This is also true of phenol.

COCAINE

Cocaine (melting point: 79°C) and similar alkaloids are extracted commercially by alkalizing the ground leaves of the coca plant with a 10% sodium carbonate solution and percolating countercurrently in large steel percolators, using kerosene or toluene. The total alkaloids are extracted from the kerosene or toluene by a process that blows them up with a 5% sulfuric acid solution in tanks. The extracted kerosene or toluene is returned to the percolators. From the sulfuric acid solution, the mixed alkaloids are precipitated by alkalizing with sodium carbonate. The precipitated crude alkaloids are slowly boiled in an 8% sulfuric acid solution for several days to split all the alkaloids to ecgonine. During the splitting, many of the organic acids, like benzoic, are partly volatilized from the kettle with the steam. Those acids that are still suspended and those that crystallize out on cooling are filtered off. The acid aqueous solution of ecgonine is neutralized with potassium carbonate and evaporated. The low-solubility potassium sulfate is filtered hot, and cooling causes the ecgonine to crystallize. After drying, it is methylated, using methanol and 92% sulfuric acid, filtered, and washed with alcohol. The methylecgonine sulfate is benzoylated to cocaine in a very vigorous reaction with benzoyl chloride in the presence of anhydrous granular potassium carbonate. The cocaine is extracted from the potassium salts with ether and removed from the ether by sulfuric acid extraction, precipitated with alkali, and crystallized from alcohol. To form the hydrochloride, an alcoholic solution is neutralized with *acid alcohol* (hydrogen chloride dissolved in absolute ethyl alcohol), and the cocaine hydrochloride is crystallized.

Because of decreased demand, cocaine is also obtained as a by-product in the preparation of a decocained extract of *Erythroxylon coca,* this being one of the principal flavors of cola beverages. The procedure is to alkalize the *Erythroxylon coca* and extract *all* the alkaloids with toluene. The decocainized leaf is dried and extracted with sherry wine to give the flavoring extract.

CODEINE

Codeine (boiling point: 250°C at 22 mmHg, melting point: 157°C, density: 1.32) can be isolated from opium and morphine (also obtained from opium). To direct this alkylation to the phenolic hydroxyl and to reduce alkylation of the tertiary nitrogen, a quaternary nitrogen-alkylating agent, phenyltrimethylammonium hydroxide, is employed. This results in yields of 90 to 93 percent codeine and some recovery of unalkylated morphine. The alkylation is carried out with the morphine dissolved in absolute alcohol in the presence of potassium ethylate. The dimethyl aniline and solvents are recovered and reused.

COKE

Coke is conventionally manufactured from coal by using the beehive process—a small batch process that tends to produce very large amounts of pollutants.

Coke is also produced from coal in the by-product ovens in which the coal charge is heated on both sides so that heat travels toward the center and thus produces shorter and more solid pieces of coke than are made in the beehive oven. Air is excluded so that no burning takes place within the oven, the heat being supplied completely from the flues on the sides. About 40 percent of the oven gas, after being stripped of its by-products, is returned and burned for the underfiring of the battery of ovens, and some is used for fuel gas locally.

Coke is also produced from petroleum, and it is the residue left by the destructive distillation of petroleum residua.

The composition of petroleum coke varies with the source of the crude oil, but in general, large amounts of high-molecular-weight complex hydrocarbons (rich in carbon but correspondingly poor in hydrogen) make up a high proportion. The solubility of petroleum coke in carbon disulfide has been reported to be as high as 50 to 80%, but this is in fact a misnomer, since the real coke is the insoluble, honeycomb material that is the end product of thermal processes.

Petroleum coke is employed for a number of purposes, but its chief use is in the manufacture of carbon electrodes for aluminum refining, which requires a high-purity carbon, low in ash and sulfur free; the volatile matter must be removed by calcining. In addition to its use as a metallurgical reducing agent, petroleum coke is employed in the manufacture of carbon brushes, silicon carbide abrasives, and structural carbon (e.g., pipes and Rashig rings), as well as calcium carbide manufacture from which acetylene is produced:

$$Coke \rightarrow CaC_2$$

$$CaC_2 + H_2O \rightarrow HC \equiv CH$$

COPPER SULFATE

Copper sulfate ($CuSO_4$) is the most important compound of copper, and is commonly known as *blue vitriol*.

Copper sulfate is manufactured by the action of sulfuric acid on cupric oxide (CuO).

Because of its poisonous nature, copper sulfate is used in the fungicide Bordeaux mixture, which is formed upon mixing copper sulfate solution with milk of lime and is added to water reservoirs to kill algae. It is also employed in electroplating and used as a mordant, germicide, and agent in engraving.

CUMENE

Cumene (*iso*-propyl benzene, boiling point: 152.4°C, density: 0.8619, flash point: 44°C) is an important intermediate in the manufacture of phenol and acetone.

Cumene is manufactured by reacting benzene with propylene over a catalyst such as a phosphoric acid derivative at 175 to 250°C and 400 to 600 psi (Fig. 1). A refinery cut of mixed propylene-propane is frequently used instead of the more expensive pure propylene. Benzene is provided in substantial excess to avoid polyalkylation. The yield is near quantitative (in excess of 90 percent) based on propylene.

$$C_6H_6 + CH_3CH=CH_2 \rightarrow C_6H_5CH(CH_3)_2$$

Excess benzene stops the reaction at the monoalkylated stage and prevents the polymerization of propylene. The cumene is separated by distillation, boiling point 153°C. Other catalysts that have been used are aluminum chloride and sulfuric acid.

Cumene is used to manufacture phenol, acetone, and α-methylstyrene

*See **Benzene.***

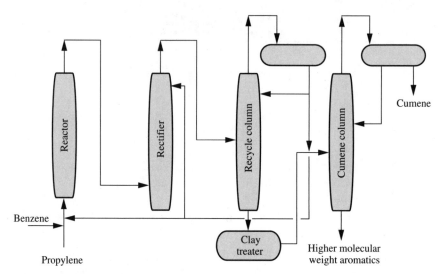

FIGURE 1 Manufacture of cumene.

CYCLOHEXANE

Cyclohexane (hexahydrobenzene, melting point: 6.5°C, boiling point: 81°C, density: 0.7791, flash point: –20°C) can be quantitatively produced from benzene by hydrogenation over either a nickel or a platinum catalyst at 210°C and 350 to 500 psi hydrogen (Fig. 1).

$$C_6H_6 + 3H_2 \rightarrow C_6H_{12}$$

Several reactors may be used in series and the yield is over 99 percent.

Cyclohexane is used for the manufacture of adipic acid and caprolactam. Adipic acid is used to manufacture nylon 6,6, and caprolactam is the monomer for nylon 6.

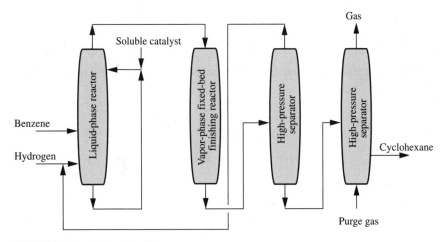

FIGURE 1   Manufacture of cyclohexane.

CYCLOHEXANOL

Cyclohexanol (melting point 25.2°C, boiling point: 161.1°C, density: 0.9493, flash point: 67.2°C) and cyclohexanone (melting point –47°C, boiling point: 156.7°C, density: 0.9478, flash point: 54°C) are made by the air oxidation of cyclohexane with a cobalt(II) naphthenate or acetate or benzoyl peroxide catalyst at 125 to 160°C (255 to 320 °F) and 50 to 250 psi.

$$2C_6H_{12} + O_2 \rightarrow C_6H_{11}-OH + C_6H_{10}=O$$

The ratio of the cyclohexanol to cyclohexanone in the product mix is 3 to 1.

The hydrogenation of phenol at elevated temperatures and pressures, in either the liquid or vapor phase and with a nickel catalyst, is also used in the manufacture of cyclohexanol.

$$C_6H_5-OH + 3H_2 \rightarrow C_6H_{11}-OH$$

Cyclohexanol is used primarily in the production of adipic acid, which is further used as a raw material in nylon 6,6 production, and in methylcyclohexanol production.

CYCLOHEXANONE

See **Cyclohexanol.**

DARVON

Darvon®, *d*-propoxyphene hydrochloride, is a synthetic, nonantipyritic, orally effective analgesic, with similar pharmacological activity and effects to codeine. Darvon is not a narcotic but can be substituted for codeine, and is useful in any condition associated with pain. Chemically, this analgesic is not analogous to codeine or to morphine.

Darvon® is a stronger analgesic than aspirin but has no antipyretic effects. It is sometimes taken in combination with aspirin and acetaminophen. It has widespread use for dental pain since aspirin is relatively ineffective, but it is not useful for deep pain.

The manufacture of Darvon (Fig. 1) starts with relatively simple chemicals, but consists of many steps :

1. Formation of a ketone from propiophenone and paraformaldehyde
2. Coupling the ketone with benzyl chloride, using the Grignard reaction (and decomposition)
3. Resolution of the optical isomers by the use of *d*-camphorsulfonic acid in acetone
4. Splitting off the *d*-camphorsulfonic acid by using ammonium hydroxide and conversion of the desired α-dextro isomer to the hydrochloride; only the dextro isomer is active as an analgesic
5. Esterification of the α-dextro isomer with propionic anhydride
6. Isolation
7. Filtration
8. Drying

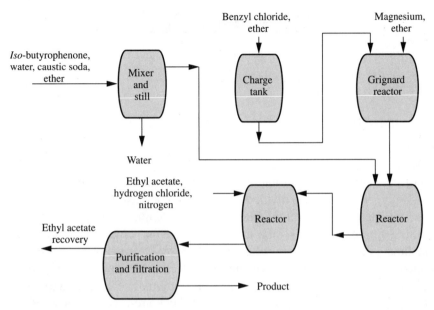

FIGURE 1 Manufacture of Darvon.

DETERGENTS

Detergents have water-attracting (hydrophilic) groups on one end of the molecule and water-repelling (hydrophobic) groups on the other.

Detergents have been divided into four main groups: anionic detergents, cationic detergents, nonionic detergents, and amphoteric detergents. The largest group consists of the anionic detergents, which are usually the sodium salts of an organic sulfate or sulfonate. Detergents can be formulated to produce a product of the desired characteristics, ranging from maximum cleaning power, maximum cleaning/unit of cost, to maximum biodegradability. Usually commercial products are a compromise of the various desirable properties.

In a process to produce a detergent containing a sulfur acid function, the alkylbenzene is introduced continuously into the sulfonator with the requisite amount of oleum (Fig. 1) to control the heat of sulfonation conversion and maintain the temperature at about 55°C. Into the sulfonated mixture is fed

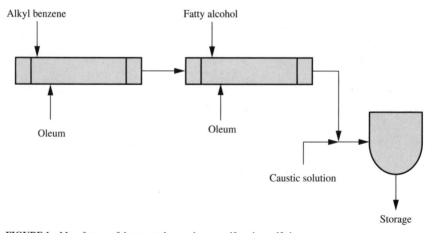

FIGURE 1   Manufacture of detergents by continuous sulfonation-sulfation.

the fatty tallow alcohol and more of the oleum. All are pumped through the sulfater, also operating on the dominant bath principle, to maintain the temperature at 50 to 55°C, thus manufacturing a mixture of surfactants. The sulfonated-sulfated product is neutralized with sodium hydroxide solution under controlled temperature to maintain fluidity of the surfactant slurry. The surfactant slurry is conducted to storage.

Detergents can be manufactured by other routes (Fig. 2), although the sulfonation-sulfation process appears to be the most popular.

Detergent builders are those chemicals that must be added to the detergent to sequester or complex the ions.

The first important commercial builder was sodium tripolyphosphate,

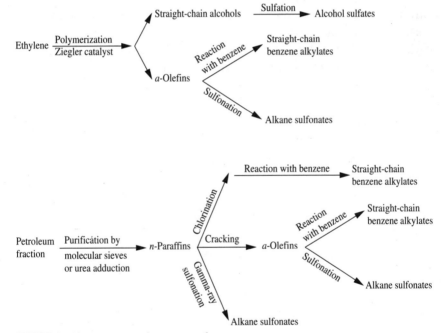

FIGURE 2   Alternate routes to detergent manufacture.

$Na_5P_3O_{10}$, first used with Tide® detergent in 1947. In the late 1960s phosphate builders came to be seen as an environmental problem. Phosphates pass unchanged through sewage works and into rivers and lakes. Since they are plant nutrients they cause blue-green algae to grow at a very fast rate on the surface, causing oxygen depletion (*eutrophication*). The search for phosphate substitutes began, and some states banned their use.

Ethylenediaminetetraacetic acid (EDTA) is a good sequestering agent but its cost is excessive. Nitrilotriacetate is effective but has been suggested to be teratogenic and carcinogenic. Sodium citrate is harmless but does not work well. Benzene polycarboxylates are expensive and are not biodegradable. Sodium carbonate is not successful in hard water areas. Commercial use of zeolites and poly-α-hydroxyacrylate is just beginning. Sodium sulfate occurs as a by-product of any sulfate or sulfonate detergent, but has limited use as a builder, as does sodium silicate. The present breakdown of builders used in detergents is sodium carbonate, 40%; sodium tripolyphosphate, 31%; sodium silicate, 9%; zeolites, 7%; sodium citrate, 3%; and other, 10%.

DIAZEPAM

*See **Benzodiazepines, Valium.***

DIAZODINITROPHENOL

*See **Explosives**.*

DIETHYLENE GLYCOL

Diethylene glycol is produced as a by-product, along with triethylene glycol, in the manufacture of ethylene glycol from hydrolysis of ethylene oxide.

$$6\overline{CH_2CH_2O} + 3H_2O \rightarrow HOCH_2CH_2OH + HOCH_2CH_2OCH_2CH_2OH +$$
$$HOCH_2CH_2OCH_2CH_2OCH_2CH_2OH$$

It is separated from the ethylene glycol and from triethylene glycol by vacuum distillation.

Diethylene glycol is used in the manufacture of polyurethane resins, unsaturated polyester resins, antifreeze blending, triethylene glycol, morpholine, and natural gas dehydration.

See *Ethylene Glycol* and *Triethylene Glycol.*

DIETHYL SULFITE

*See **Sulfurous Acid.***

DIHYDROXYACETONE

Dihydroxyacetone ($HOCH_2COCH_2OH$, melting point: 89°C) is made by the action of sorbose bacterium fermentation of glycerin ($HOCH_2CHOHCH_2OH$).

Dihydroxyacetone is an ingredient of suntan lotion that creates an artificial tan. It is also valuable as a chemical intermediate and as a catalyst in butadiene-styrene polymerization.

DIMETHYL SULFITE

*See **Sulfurous Acid.***

DIMETHYL TEREPHTHALATE

Dimethyl terephthalate (melting point: 141°C) is prepared by oxidation of *p*-xylene and subsequent esterification with methyl alcohol (Fig. 1).

$$CH_3C_6H_4CH_3 + [O] \rightarrow HOOCC_6H_4COOH$$

$$HOOCC_6H_4COOH + 2CH_3OH \rightarrow CH_3OOCC_6H_4COOCH_3$$

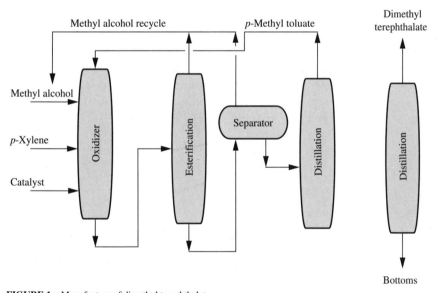

FIGURE 1 Manufacture of dimethyl terephthalate.

2,4- AND 2,6- DINITROTOLUENE

The dinitrotoluenes [$CH_3C_6H_3(NO_2)_2$] are manufactured by the nitration of toluene that gives a mixture containing 65 to 80% of the 2,4-dinitro derivative and 20 to 35% of the 2,6-dinitro compound.

$$2C_6H_5CH_3 + 4HNO_3 + 4H^+ \rightarrow 2{,}4\ (NO_2)_2C_6H_3CH_3 \\ + 2{,}6\ (NO_2)_2C_6H_3CH_3 + 4H_2O$$

If the pure 2,4-compound is required, mononitration of toluene followed by separation of pure *p*-nitrotoluene from the ortho isomer, and then further nitration of *p*-nitrotoluene gives the pure 2,4-dinitro isomer.

Nearly all the dinitrotoluenes are hydrogenated to diamines, which are converted into diisocyanates to give toluene diisocyanate, a monomer for polyurethane. A small amount of the dinitrotoluenes is further nitrated to the explosive 2,4,6-trinitrotoluene (TNT).

DIPHENYL ETHER

Diphenyl ether (diphenyl oxide) is obtained as a by-product in the manufacture of phenol from chlorobenzene and caustic soda.

$$2C_6H_5OH \rightarrow C_6H_5OC_6H_5 + H_2O$$

Diphenyl ether is used in the soap and perfume industries because of its great stability and strong geranium-type odor.

DYAZIDE

Dyazide®, a diuretic, contains both triamterene and hydrochlorothiazide. Triamterene is a diuretic and is known to increase sodium and chloride ion excretion but not potassium ion.

A number of thiazides can be synthesized from appropriate sulfonamides by cyclization with dehydration. Conversion to hydrothiazides increases their activity by a factor of 10.

Dyazide is used in conjunction with a hydrothiazide, which is an excellent diuretic but also gives significant loss of potassium and bicarbonate ions. If the triamterene were not included, potassium chloride would have to be added to the diet. Hydrochlorothiazide is an antihypertensive agent as well but, unlike other antihypertensives, it lowers blood pressure only when it is too high, and not in normotensive individuals.

DYES

A dye is a colored chemical that can impart color to a material or other body on a reasonably permanent basis. Most dyes contain considerable unsaturation, and some part of the dye is usually in the form of aromatic rings with nitrogen unsaturation of several types common to many dyes. The quinoid structure appears frequently, although other dye families are available.

It is possible to classify dyes by using the *Color Index* that classifies dyes according to a dual system. An assigned number defines the chemical class and a generic name identifies the usage of application. However, it is convenient to use the application classification used by the U.S. International Trade Commission for application classes:

1. Acid dyes
2. Azoic dyes
3. Basic dyes
4. Direct dyes
5. Disperse dyes,
6. Fiber-reactive dyes
7. Fluorescent brightening agents
8. Food drug and cosmetic colors
9. Mordant dyes
10. Solvent dyes,
11. Sulfur dyes
12. Vat dyes

Acid dyes derive their name from their insolubility in acid baths. They are used for dyeing protein fibers such as wool, silk, and nylon; also leather and paper.

Azoic dyes, the "ice colors," are made on the fiber by coupling diazotized materials while in contact with the fibers. Low temperature keeps the diazonium compound from decomposing until ready to couple.

Basic dyes are mostly amino and substituted amino compounds soluble in acid and made insoluble by making the solution basic.

Direct dyes are used to dye cotton directly, that is, without the addition of a mordant.

Disperse dyes are applied to difficult-to-dye materials as very finely divided materials that are adsorbed onto the fibers, with which they then form a solid solution,

Fiber-reactive dyes react to form a covalent link between dye and the cellulosic fiber which they are customarily used to dye.

Fluorescent brightening agents are used to provide greater brilliance than can be obtained with soap, textiles, plastics, paper, and detergents.

Food, drug, and cosmetic colors consist of a carefully controlled group of regulated materials. Purity and safety are rigidly monitored; some dyes are *listed* and some are *certified.*

Mordant dyes (and lakes) are dyes combined with metallic salts (mordant means bitter) to form highly insoluble colored materials called *lakes.*

Solvent dyes are dyes that are soluble in alcohols, chlorinated hydrocarbon solvents, or liquid ammonia, and there appears to be considerable promise in dyeing the difficult-to-dye synthetics, polyesters, polyacrylates, and triacetates, from such solutions.

Sulfur dyes (sulfide dyes) are a large, low-cost group of dyes that produce dull shades on cotton. Sulfur dyes are usually colorless when in the reduced form in a sodium sulfide bath but gain color on oxidation.

Vat dyes have complex chemical structures that are (for the most part) derivatives of anthraquinone or indanthrene.

DYNAMITE

Dynamite was originally made by absorbing nitroglycerin into kieselguhr, a type of clay. Modern dynamite generally includes wood flour, ammonium nitrate, or sodium nitrate to absorb the nitroglycerin. Such a mixture is easy to handle and can be made to contain as much as 75% nitroglycerin and yet retain its solid form.

Because of the demand for a nonfreezing dynamite for use in cold weather, dynamites containing other materials designed to lower the freezing point of the mixture are used, for example, glycol dinitrate ($CH_2ONO_2CH_2ONO_2$). Such nonfreezing dynamites have potential as *straight* dynamite. Nitrocellulose can be gelatinized by nitroglycerin, and the resultant firm gel is commonly known as gelatin dynamite.

The ability to act as a combination plasticizer and explosive makes nitroglycerin and the similar diethylene glycol dinitrate (DEGN) useful in plastic explosives and smokeless powder manufacture.

Almost without exception the nitro compounds and nitric acid esters used as explosives are toxic. The degree of toxicity varies widely with the material in question, but most are capable of causing acute distress if taken orally. Nitroglycerin has a small medical use as a vasodilator.

*See **Nitroglycerin.***

EPOXY RESINS

The most common epoxy resins are formed by the reaction of bisphenol A with epichlorohydrin.

Epoxy resins are intermediates for the production of other materials and, as such, must be cured, or cross-linked, to yield a useful resin. Cross-linking occurs by the opening of the epoxide ring by addition of a curing agent that must have active hydrogen atoms. Amines, acid anhydrides, and mercaptans are the most usual compounds used as curing agents.

Depending upon molecular weight, epoxy resins have a great many uses, ranging from adhesives to can and drum coatings. They have excellent chemical resistance, particularly to alkalis, very low shrinkage on cure, excellent adhesion and electrical insulating properties, and ability to cure over a wide range of temperatures.

See Bisphenol A.

ERYTHROMYCIN

Erythromycin is, like penicillin, isolated by solvent-extraction methods. It is an organic base, and extractable with amyl acetate or other organic solvents under basic conditions rather than the acidic ones that favor penicillin extraction.

ETHANE

Ethane (C_2H_6, melting point: $-172°C$, boiling point: $-88.6°C$) is a colorless, odorless gas that has a very slight solubility in water but is moderately soluble in alcohol.

Ethane occurs in natural gas, from which it is isolated. Ethane is among the chemically less reactive organic substances. However, ethane reacts with chlorine and bromine to form substitution compounds. Ethyl iodide, bromide, or chlorides are preferably made by reaction with ethyl alcohol and the appropriate phosphorus halide. Important ethane derivatives, by successive oxidation, are ethyl alcohol, acetaldehyde, and acetic acid. Ethane can also be used for the production of aromatics by pyrolysis (Fig. 1).

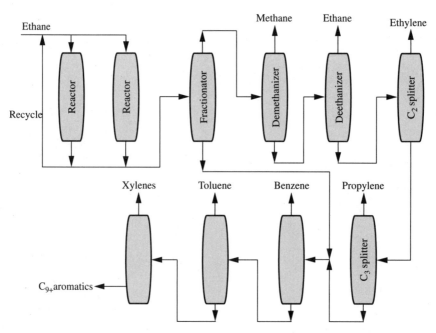

FIGURE 1   Manufacture of aromatics from ethane.

ETHANOLAMINES

The ethanolamines monoethanolamine (MEA, $HOCH_2CH_2NH_2$, melting point: 10.5°C, boiling point: 171°C, density: 1.018), diethanolamine [DEA, $(HOCH_2CH_2)_2NH$, melting point: 28.0°C, boiling point: 270°C, density: 1.019], and triethanolamine [TEA, $(HOCH_2CH_2)_3N$, melting point: 21.2°C, boiling point: 360°C, density: 1.126] are manufactured by reacting ethylene oxide and excess ammonia, followed by separation of unreacted ammonia and the three ethanolamines. The proportion of the three products depends on reaction conditions.

Mono- and triethanolamine are miscible with water or alcohol in all proportions and is only slightly soluble in ether. Diethanolamine will dissolve in water, is very soluble in alcohol, and is only slightly soluble in ether. All of the compounds are clear, viscous liquids at standard conditions and white crystalline solids when frozen. They have a relatively low toxicity.

In early processes, the ethanolamines were manufactured by reacting ethylene chlorohydrin ($ClCH_2CH_2OH$) with ammonia (NH_3). Current processes react ethylene oxide ($\overline{CH_2CH_2O}$) with ammonia usually in aqueous solution. The ratio of mono-, di-, and triethanolamines varies in accordance with the amount of ammonia present. This is controlled by the quantities of monoethanolamine and diethanolamine recycled.

Higher ammonia-ethylene oxide ratios favor high yields of diethanolamine and triethanolamine, whereas lower ratios are used where maximum production of monoethanolamine is desired. The reaction is noncatalytic. The pressure is moderate, just sufficient to prevent vaporization of components in the reactor. The bulk of the water produced in the reaction is removed by subsequent evaporation. The dehydrated ethanolamines then proceed to a further drying column, after which they are separated in a series of fractionating columns, not difficult because of the comparatively wide separation of their boiling points.

Industrially, the ethanolamines are important because they form numerous derivatives, notably with fatty acids, soaps, esters, amides, and esteramides, and they have an exceptional ability for scrubbing acidic compounds out of gases. Monoethanolamine, for example, will effectively remove hydrogen sulfide (H_2S) from hydrocarbon gases, and carbon dioxide can also be removed from process streams and, where desired, heating the absorptive solutions can regenerate the carbon dioxide.

The soaps of the ethanolamines are extensively used in textile treating agents, in shampoos, and emulsifiers. The fatty acid amides of diethanolamine are applied as builders in heavy-duty detergents, particularly those in which alkylaryl sulfonates are the surfactant ingredients. The use of triethanolamine in photographic developing baths promotes fine grain structure in the film when developed.

Ethanolamine also is used as a humectant and plasticizing agent for textiles, glues, and leather coatings and as a softening agent for numerous materials. Morpholine is an important derivative.

See Amination.

ETHER

Diethyl ether (ether, $C_2H_5OC_2H_5$, melting point: $-116.3°C$, boiling point: $34.6°C$, density: 0.708) is slightly soluble in water (1 volume in 10 volumes of water) and is miscible with alcohol in all proportions.

The long-used manufacturing procedure for ether has been the dehydration of alcohol (denatured with ether) by sulfuric acid. The anesthetic ether is especially purified and packaged. Ether is also supplied as a by-product from the manufacture of alcohol from ethylene.

Ether is a valuable solvent for organic chemicals. At one time ether was the major anesthetic, for which it must be scrupulously pure. In addition to various side effects that may result from the use of ether as an anesthetic, it is a definite hazard in the operating room because of its explosive properties, particularly in enriched oxygen atmospheres.

ETHYL ACETATE

Ethyl acetate (boiling point: 77.1, density: 0.9005, flash point: –4°C) is manufactured from ethyl alcohol and acetic acid.

$$CH_3COOH + C_2H_5OH \rightarrow CH_3COOC_2H_5$$

A more recent process allows the manufacture of ethyl acetate from ethyl alcohol without the use of acetic acid or any other cofeedstocks. In the process (Fig. 1), ethyl alcohol is heated and passed through a catalytic dehydrogenation reactor where part of the ethyl alcohol is dehydrogenated to form ethyl acetate and hydrogen.

$$C_2H_5OH \rightarrow [CH_3CHO] + H_2$$

$$[CH_3CHO] + C_2H_5OH \rightarrow CH_3COOC_2H_5 + H_2$$

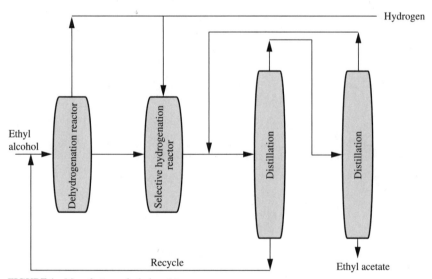

FIGURE 1   Manufacture of ethyl acetate.

ETHYL ALCOHOL

Ethyl alcohol (ethanol, freezing point: $-114.1°C$, boiling point: $78.3°C$, density: 0.7893, flash point: $14°C$) is also named, industrial alcohol, grain alcohol, and alcohol. Ethyl alcohol is miscible in all proportions with water or with ether. When ignited, ethyl alcohol burns in air with a pale blue, transparent flame, producing water and carbon dioxide. The vapor forms an explosive mixture with air and is used in some internal combustion engines under compression as a fuel; such mixtures are frequently referred to as gasohol.

Much of the ethanol produced in the past has been manufactured by routes such as by the fermentation of sugars. Waste syrup (molasses) after crystallization of sugar from sugar cane processing, containing mostly sucrose, can be treated with enzymes from yeast to give ethyl alcohol and carbon dioxide. The enzymes require 28 to 72 hours at 20 to $38°C$ and give a 90 percent yield. This process is not used to any great extent for the current production of industrial alcohol.

$$\underset{\text{Sucrose}}{C_{12}H_{22}O_{11}} + H_2O \rightarrow \underset{\text{Glucose and/or fructose}}{2C_6H_{12}O_6}$$

$$C_6H_{12}O_6 \rightarrow 2CH_3CH_2OH + 2CO_2$$

The next method of manufacture, the esterification/hydrolysis of ethylene, was another method of choice, and an important side product in this electrophilic addition to ethylene was diethyl ether.

$$CH_2{=}CH_2 + H_2SO_4 \rightarrow CH_2CH_2OSO_3H$$

$$CH_2CH_2OSO_3H + H_2O \rightarrow CH_3CH_2OH + H_2SO_4$$

The current process of choice involves the direct hydration of ethylene with a catalytic amount of phosphoric acid (Fig. 1). Temperatures average 300 to $400°C$ with 1000 psi.

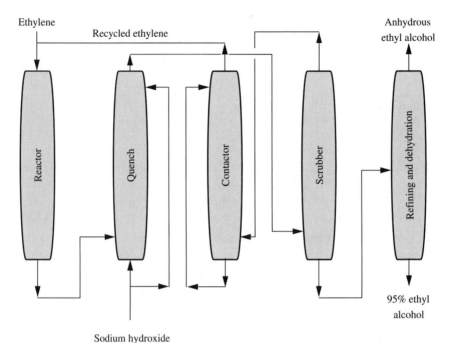

FIGURE 1   Manufacture of ethyl alcohol.

$$CH_2=CH_2 + H_2O \rightarrow CH_2CH_2OH$$

Only 4 percent of the ethylene is converted to alcohol per pass, but this cyclic process eventually gives a net yield of 97 percent.

In this direct hydration process, a supported acid catalyst usually is used. Important factors affecting the conversion include temperature, pressure, the water/ethylene ratio, and the purity of the ethylene. Further, some by-products are formed by other reactions taking place, a primary side reaction being the dehydration of ethyl alcohol into diethyl ether:

$$2C_2H_5OH \rightarrow (C_2H_5)_2O + H_2O$$

To overcome these problems, a large recycle volume of unconverted ethylene usually is required. The process usually consists of a reaction section in which crude ethyl alcohol is formed, a purification section with a product of 95% (volume) ethyl alcohol, and a dehydration section, which produces high-purity ethyl alcohol free of water. For many industrial uses, the 95%-purity product from the purification section suffices.

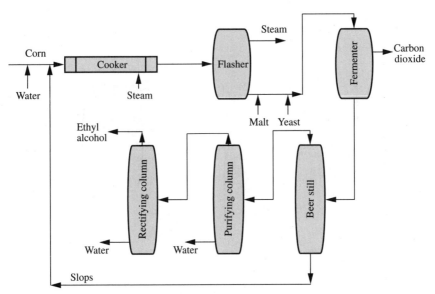

FIGURE 2   Production of ethyl alcohol by fermentation.

The fermentation process (Fig. 2) is in use again to produce ethyl alcohol for use in gasohol, an automobile fuel that is a simple mixture of 90% gasoline and 10% alcohol claimed to increase mileage.

Ethyl alcohol is also manufactured from a cellulose source (such as wood) (Fig. 3) but no matter what the process, it is often necessary to dehydrate the product for the production of pure alcohol (Fig. 4).

Ethyl alcohol is also produced by the vapor phase reduction of acetaldehyde in the presence of metal catalysts.

$$CH_3CH{=}O \;\rightarrow\; CH_3CH_2OH$$

Suitable catalysts for the process are supported nickel and copper oxide.

Anhydrous ethyl alcohol is made from the constant boiling mixture with water (95.6% ethyl alcohol by weight) by heating with a substance such as calcium oxide, which reacts with water and not with alcohol, and then distilling, or by distilling with a volatile liquid, such as benzene (boiling point: 79.6°C) which forms a constant low-boiling mixture with water and alcohol (boiling point: 64.9°C), so that water is removed from the main portion of the alcohol; after which alcohol plus benzene distills over (boiling point: 78.5°C).

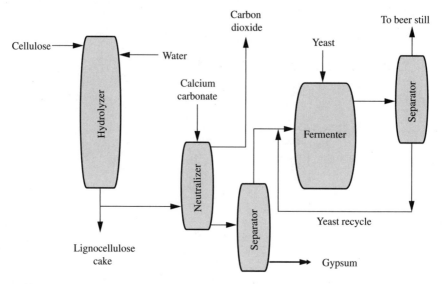

FIGURE 3   Production of ethyl alcohol from cellulose (wood)

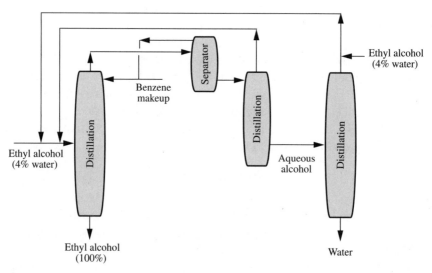

FIGURE 4   Dehydration of crude ethyl alcohol.

Most alcohol is sold as 95% ethanol–5% water since it forms an azeotrope at that temperature. To obtain absolute alcohol, a third component, such as benzene, must be added during the distillation. This tertiary azeotrope carries over the water and leaves the pure alcohol behind. Common industrial alcohol is *denatured,* and additives are purposely included to make it nondrinkable and therefore not subject to the high taxes of the alcoholic beverage industry.

The use and production values for ethanol do not include that amount produced for most alcoholic beverages.

U.S. proof was the term originally applied to a test of strong alcoholic beverages done by pouring the sample onto gunpowder and lighting it. Ignition of the gunpowder was considered proof that the beverage was not diluted. A value of 100 proof now refers to 50% alcohol by volume (42.5% by weight). Thus 190 proof is 95% alcohol by volume (92.4% by weight).

Industrial uses of ethanol include solvents (especially for toiletries and cosmetics, coatings and inks, and detergents and household cleaners), and chemical intermediates (especially ethyl acrylate, vinegar, ethylamines, and glycol ethers). Most corn fermentation alcohol is used in fuel, industrial solvents and chemicals, and beverages.

ETHYLBENZENE

Ethylbenzene (boiling point: 136°C, density: 0.8672, flash point: 21°C) is a colorless liquid that is manufactured from benzene and ethylene by several modifications of the older mixed liquid-gas reaction system using aluminum chloride as a catalyst (Friedel-Crafts reaction). The reaction takes place in the gas phase over a fixed-bed unit at $370°C$ under a pressure of 1450 to 2850 kPa. Unchanged and polyethylated materials are recirculated, making a yield of 98 percent possible. The catalyst operates several days before requiring regeneration.

$$C_6H_6 + CH_2=CH_2 \text{ (+AlCl}_3/\text{HCl or zeolite)} \rightarrow C_6H_5CH_2CH_3 \text{ (+AlCl}_3/\text{HCl or zeolite)}$$

Excess benzene is used if the formation of di- and trimethylbenzenes is to be avoided or minimized. The benzene is recycled.

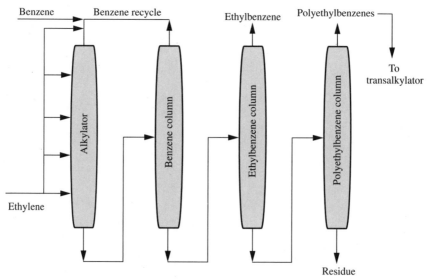

FIGURE 1 Manufacture of ethylbenzene.

In the more modern process (Fig. 1) the reaction takes place in the liquid phase, using a zeolite catalyst and cycle lengths in excess of 3 years are expected for the catalyst.

A vapor-phase method with boron trifluoride, phosphoric acid, or alumina-silica as catalyst has given away to a liquid-phase reaction with aluminum chloride at 90°C and atmospheric pressure. A zeolite catalyst at 420°C and 175 to 300 psi in the gas phase is also available.

Despite the elaborate separations required, including washing with caustic and water and three distillation columns, the overall yield of ethylbenzene is 98 percent. Ethylbenzene is used predominantly ($C_6H_5CH=CH_2$) manufacture; a minor amount is used as a solvent.

See **Benzene and Xylenes.**

ETHYLENE

Ethylene (boiling point: −103.7°C, flash point: −136.1°C, ignition temperature: 450°C) is a colorless, flammable gas with a faint, pleasant odor that can be isolated from refinery gas (Fig. 1), but the majority of the ethylene (ethene) is produced by the thermal cracking of petroleum feedstocks (hydrocarbons) at high temperatures with no catalyst. In contrast to the catalytic cracking used by the petroleum industry to obtain high yields of gasoline, thermal cracking is used since it yields larger percentages of ethylene, propylene, and butylenes. Naphtha and gas oil fractions from petroleum can also be used as feedstock for ethylene manufacture.

Steam cracking (Fig. 2), consists of a furnace in which the cracking takes place is at 815 to 870°C (1500 to 1600°F). As many as 6 to 20 furnaces are in parallel to increase ethylene production. Steam is used as a diluent to inhibit coking in the tubes and to increase the percentage of ethylene formed. The amount of steam changes with the molecular weight of the hydrocarbon feedstock and varies from 0.3 kg steam/kg ethane to 0.9

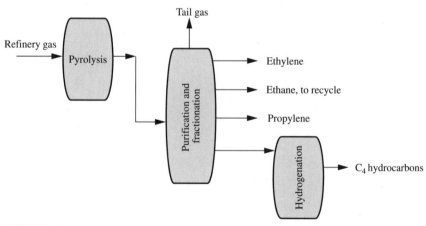

FIGURE 1   Production of ethylene from refinery gas.

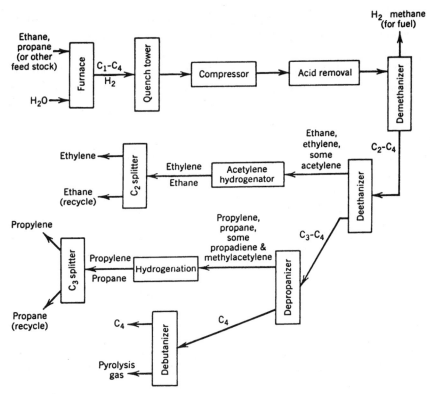

FIGURE 2   Olefin manufacture by thermal cracking.

kg steam/kg gas oil. Contact time is 1 second or less in the furnace. The exit gases are immediately cooled in the quench tower then placed under 500 psi pressure by a compressor. Monoethanolamine or caustic soda solution is used to remove hydrogen sulfide and carbon dioxide. The demethanizer, deethanizer, and debutanizer are fractionating columns that separate the lighter and heavier compounds from each other.

Traces of other unsaturated triple bonds are removed by catalytic hydrogenation with a palladium catalyst in both the C_2 and C_3 stream. Cumulated double bonds are also hydrogenated in the C_3 fraction. These are more reactive in hydrogenation than ethylene or propylene. The C_2 and C_3 fractionators are distillation columns that can be as high as 200 ft.

Lower-molecular-weight feedstocks, such as ethane and propane, give a high percentage of ethylene; higher-molecular-weight feedstocks, such as naphtha and gas oil, are used if propylene is required in significant quantities.

Ethylene is sold from 95% purity (technical) to 99.9% purity. It can be transported by pipeline or by tank car. Smaller amounts come in 100-lb cylinders.

Ethylene is used in the manufacture of a wide range of chemicals (Fig. 3) including polyethylene (Figs. 4 and 5), styrene, alcohols, ethylene oxide, and vinyl chloride.

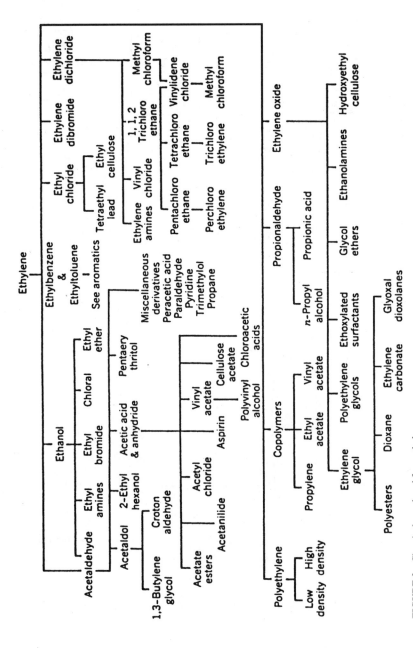

FIGURE 3 Chemicals produced from ethylene.

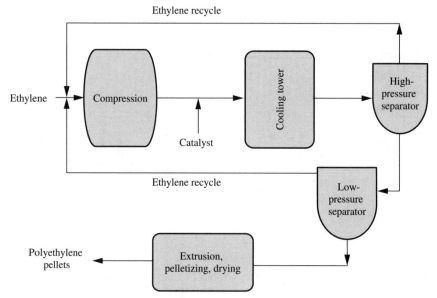

FIGURE 4   Manufacture of low-density polyethylene

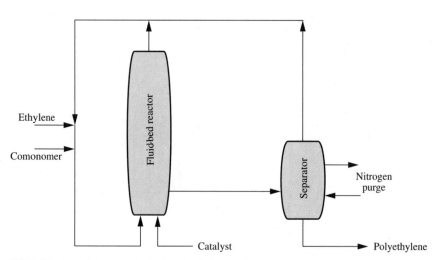

FIGURE 5   Manufacture of high-density polyethylene.

ETHYLENE DICHLORIDE

Ethylene dichloride (1,2-dichloroethane, CH$_2$ClCH$_2$Cl), a colorless toxic liquid (boiling point 84°C, density: 1.2560, flash point: 13°C) is manufactured by two methods.

The classical method for the manufacture of ethylene dichloride is the addition of chlorine to the double bond of ethylene (Fig. 1).

$$CH_2=CH_2 + Cl_2 \rightarrow CH_2ClCH_2Cl$$

Ferric chloride is often used as the catalyst. The yield is high (96 to 98%, based on ethylene) and the reaction can be carried out in the vapor phase or in the liquid phase at varying temperatures and the product is easily purified by fractional distillation.

However, when chlorine is added to olefins such as ethylene, many and mixed derivatives are formed. In addition to ethylene dichloride, dichloroethylene, trichloroethylene, tetrachloroethane, and chloromethanes form. Other halogens produce similar mixtures. With good reaction control, high yields of the desired product are possible.

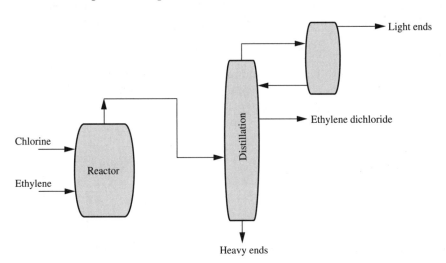

FIGURE 1 Manufacture of ethylene dichloride by direct chlorination.

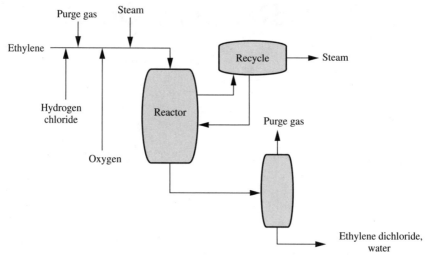

FIGURE 2   Manufacture of ethylene dichloride by oxychlorination.

In contrast to this direct chlorination there is the oxychlorination of ethylene using hydrogen chloride and oxygen (Fig. 2).

$$2CH_2{=}CH_2 + 4HCl + O_2 \rightarrow 2CH_2ClCH_2Cl + 2H_2O$$

The catalysts for this reaction are cupric chloride ($CuCl_2$), potassium chloride (KCl), and alumina (Al_2O_3) or silica (SiO_2).

Most of the ethylene dichloride manufactured is converted into vinyl chloride by eliminating a mole of hydrogen chloride (HCl) that can then be recycled and used to make more ethylene dichloride by oxychlorination.

$$CH_2ClCH_2Cl \rightarrow CH_2{=}CHCl + HCl$$

Ethylene is also used to manufacture perchloroethylene ($CCl_2{=}CCl_2$), methyl chloroform (CH_3CCl_3), vinylidene chloride ($CH_2{=}CCl_2$), and ethylamines (e.g., $H_2NCH_2CH_2NH_2$).

ETHYLENE GLYCOL

The primary manufacturing method of making ethylene glycol (ethane-1,2-diol, boiling point: 197.6°C, density: 1.1155, flash point: 127°C) is from acid or thermal-catalyzed hydration and ring opening of the oxide.

$$\overline{CH_2CH_2O} + H_2O \rightarrow HOCH_2CH_2O$$
ethylene glycol

$$2\overline{CH_2CH_2O} + H_2O \rightarrow HOCH_2CH_2OCH_2CH_2OH$$
diethylene glycol

$$3\overline{CH_2CH_2O} + H_2O \rightarrow HOCH_2CH_2OCH_2CH_2OCH_2CH_2O$$
triethylene glycol

In the process (Fig. 1), either a 0.5 to 1.0% sulfuric acid (H_2SO_4) catalyst is used at 50 to 70°C for 30 minutes or, in the absence of the acid, a temperature of 195°C and 185 psi for 1 hour will form the diol. A 90 percent yield is realized when the ethylene oxide/water molar ratio is 1:5-8. The advantage of the acid-catalyzed reaction is no high pressure; the thermal reaction however needs no corrosion resistance and no acid separation step.

The crude glycols are dehydrated and then recovered individually as highly pure overhead streams from a series of vacuum-operated purification columns. Ethylene glycol (boiling point: 197.6°C) is readily vacuum distilled and separated from the diethylene glycol (boiling point: 246°C, density: 1.118, flash point: 124°C) and triethylene glycol (boiling point: 288°C, density: 1.1274, flash point: 177°C).

Ethylene glycol is traditionally associated with use as permanent-type antifreeze for internal-combustion engine cooling systems. Other uses include the production of polyesters for fibers, films, and coatings, in hydraulic fluids, in the manufacture of low freezing-point explosives, glycol ethers, and deicing solutions.

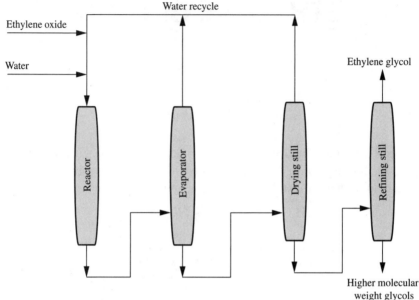

FIGURE 1   Manufacture of ethylene glycol.

Di and triethylene glycols are important co-products usually produced in the manufacture of ethylene glycol.

Diethylene glycol ($HOCH_2CH_2OCH_2CH_2OH$), is used in the production of unsaturated polyester resins and polyester polyols for polyurethane-resin manufacture, as well as in the textile industry as a conditioning agent and lubricant for numerous synthetic and natural fibers. It is also used as an extraction solvent in petroleum processing, as a desiccant in natural gas processing, and in the manufacture of some plasticizers and surfactants.

Triethylene glycol ($HOCH_2CH_2OCH_2CH_2OCH_2CH_2OH$) finds principal use in the dehydration of natural gas and as a humectant.

ETHYLENE OXIDE

Ethylene oxide (freezing point: –111.7°C, boiling point: 10.4°C, flash point: <18°C) is a colorless gas that condenses at low temperature into a mobile liquid. Ethylene oxide is miscible in all proportions with water or alcohol and is very soluble in ether. Ethylene oxide is slowly decomposed by water at standard conditions, converting into ethylene glycol ($HOCH_2CH_2OH$).

For many years the manufacture of ethylene oxide was carried out by chlorohydrin formation followed by dehydrochlorination to the epoxide.

$$CH_2{=}CH_2 + Cl_2 + H_2O \rightarrow ClCH_2CH_2OH + HCl$$

$$ClCH_2CH_2OH \rightarrow \overline{CH_2CH_2O} + HCl$$

Although the chlorohydrin route is still used to convert propylene to propylene oxide, a more efficient air epoxidation of ethylene in the presence of a silver catalyst is used and involves a direct oxidation method (Fig. 1).

$$2CH_2{=}CH_2 + O_2 \rightarrow 2\overline{CH_2CH_2O}$$

The yield is approximately 70 percent of the theoretical. For maximum yield, very careful temperature control is required, the yield dropping as the temperature climbs. Excessive oxidation to carbon dioxide and water can occur, so precise temperature control at 270 to 290°C and pressures of 120 to 300 psi with a 1-second contact time on the catalyst are necessary. Tubular reactors containing several thousand tubes (20 to 50 mm diameter) are used. Even though metallic silver is placed in the reactor, the actual catalyst is silver oxide under the conditions of the reaction.

Ethylene oxide is used for manufacture of ethylene glycol, the latter being an antifreeze compound as well as a raw material for production of polyethylene terephthalate used in the manufacture of polyester fibers; for preparation of surfactants; for the manufacture of ethanolamines; for production of ethylene glycols used in plasticizers, solvents, and lubricants; and for making glycol ethers used as jet-fuel additives and solvents.

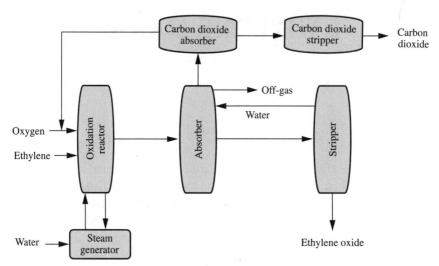

FIGURE 1   Manufacture of ethylene oxide.

Diethylene glycol and triethylene glycol (DEG and TEG) are produced as by-products of ethylene glycol and are used in polyurethane and unsaturated polyester resins and in the drying of natural gas. Diethylene glycol also used in antifreeze and in the synthesis of morpholine, a solvent, corrosion inhibitor, antioxidant, and pharmaceutical intermediate.

*See **Ethylene Glycol, Diethylene Glycol, Triethylene Glycol.***

ETHYLHEXANOL

2-ethylhexanol is produced by aldol condensation of butylaldehyde followed by reduction.

$$CH_3CH_2CH_2CH=O + H_2 \rightarrow CH_3CH_2CH_2CH_2CH(CH_2CH_3)CH_2OH$$

2-ethylhexanol is used for the production of plasticizers, especially dioctyl phthalate, dioctyl adipate, and trioctyl trimellitate, as well as in the production of 2-ethylhexyl acrylate for adhesives and coatings, and for the production of 2-ethylhexylnitrate.

ETHYL VINYL ETHER

*See **Vinyl Ethers.***

EXPLOSIVE D

See Ammonium Picrate.

EXPLOSIVES

An explosive is a material that, under the influence of thermal or mechanical shock, decomposes rapidly and spontaneously with the evolution of a great deal of heat and gas. The hot gases cause extremely high pressure if the explosive is set off in a confined space.

Initiating explosives (primary explosives) are materials that are quite shock and heat sensitive and that can be made to explode by the application of a spark, flame, friction, or heat source of appropriate magnitude.

Primary explosives include mercury fulminate [$Hg(ONC)_2$, melting point: 160°C with explosion, density: 4.2], lead azide [$Pb(N_3)_2$, density: 4.0], basic lead styphnate (lead trinitroresorcinate), diazodinitrophenol, and tetrazine (a complex conjugated nitrogen compound, melting point: 140 to 160°C with explosion). Most priming compositions consist of mixtures of primary explosives, fuels, and oxidants.

Booster high explosives (energy amplifiers) are materials that are insensitive to both mechanical shock and flame but that explode with great violence when set off by an explosive shock, such as that obtained by detonating a small amount of a primary explosive in contact with the high explosive.

Boosters such as RDX (cyclotrimethylenetrinitramine, melting point: 204°C, density: 1.80), PETN (pentaerythritoltetranitrate, melting point: 143°C, density: 1.78), and TETRYL (2,4,6-trinitrophenyl methyl nitramine, melting point: 129.5°C, density: 1.73) are extremely important chemicals.

Blasting agents are powerful explosive agents that cannot be detonated by means of a blasting cap when unconfined and are, therefore, very safe to handle. A powerful booster is needed to start detonation. Blasting agents are usually ammonium nitrate (melting point: 169.6°C, density: 1.725) mixtures sensitized with nonexplosive fuels such as oil or wax.

Slurry explosives are ammonium nitrate mixtures that frequently contain another oxidizer as well as a fuel dispersed in a fluid medium that, among other functions, controls the rheology of the gel-slurry. Like blasting agents, slurry explosives are usually ammonium nitrate mixtures sensitized with nonexplosive fuels such as oil or wax.

FERRIC OXIDE

Ferric oxide (Fe_2O_3) occurs naturally and is often used as a red pigment employed in paints and primers, as well as in rubber formulation. Because of durability, the iron oxide pigments are used in barn and freight car paints. The synthetic pigment is made by heating iron sulfate.

Venetian red is a mixture of ferric oxide with up to an equal amount of the pigment extender, calcium sulfate. This pigment is manufactured by heating ferrous sulfate with quicklime in a furnace. Venetian red is a permanent and inert pigment, particularly on wood. The calcium sulfate content, which furnishes corrosion-stimulating sulfate ions, disqualifies this pigment for use on iron.

Indian red is a naturally occurring mineral whose ferric oxide content may vary from 80 to 95%, the remainder being clay and silica. It is made by grinding hematite and floating off the fines for use.

FERROCYANIDE BLUE

Ferrocyanide blue occurs in various shades known as Prussian blue, Chinese blue, muon blue, bronze blue, Antwerp blue, and Tumbull's blue. These names have lost much of their original differentiation; the more general term *iron blues* is preferred.

These pigments are manufactured by treating ferrous sulfate ($FeSO_4$) solutions (sometimes in the presence of ammonium sulfate) with sodium ferrocyanide, giving a white ferrous ferrocyanide, which is then oxidized to ferric ferrocyanide, $Fe_4[Fe(CN)_6]$, or to $Fe(NH_4)[Fe(CN)_6]$ by different reagents such as potassium chlorate, bleaching powder, and potassium dichromate. The colloidal pigment is washed and allowed to settle to enhance separation, since filtration of the colloidal solid is difficult.

Iron blues possess very high tinting strength and good color performance and the relative transparency of these pigments is an advantage in dip-coating foils and bright metal objects, and for colored granules for asphalt shingles. They cannot be used in water-based paints because of their poor alkali resistance.

FERTILIZERS

Fertilizers provide the primary nutrients (nitrogen, phosphorus, and potassium) for vegetation and are manufactured by a variety of processes (Fig. 1) and from a variety of raw materials (Fig. 2). The usual sources of nitrogen are ammonia, ammonium nitrate, urea, and ammonium sulfate. Phosphorus is obtained from phosphoric acid or phosphate rock and potassium is available from mined potassium chloride or potassium sulfate or it is obtained from brine.

Raw materials	Primary chemicals	Derived chemicals
Air	Ammonia	Ammonium sulfate
Hydrogen	Ammonia	Urea
		Ammonium nitrate
Phosphate rock	Phosphoric acids	Ammonium phosphate
		Superphosphates

FIGURE 1   Routes for the manufacture of fertilizers.

Raw materials	Primary product	Secondary product
Wood	Ammonia	Urea
Lignite		Ammonium nitrate
Coal		Ammonium sulfate
Hydrogen		Ammonium chloride
Coke oven gas		Ammonium phosphates
Natural gas		
Liquefied petroleum gas		
Refinery gas		
Naphtha		
Fuel oil		
Bunker C oil		

FIGURE 2   Raw materials used for fertilizer manufacture.

Fertilizers may contain all three primary nutrients, in which case they are called *mixed fertilizers*, or they may contain only one active ingredient, called *direct application fertilizers*. The advantage of using mixed fertilizers is that they contain all three primary nutrients—nitrogen, phosphorus, and potassium —and require a smaller number of applications. They can be liquids or solids. The overall percentage of the three nutrients must always be stated on the container. The grade designation is $%N-%P_2O_5-%K_2O$ and is commonly referred to as the *nitrogen-phosphorus-potassium (NPK) value.*

Phosphorus-based fertilizers usually are produced from wet-process phosphoric acid or directly from phosphate rock. Normal superphosphate, triple or concentrated superphospate, and ammonium phosphate are the three common types used. Normal or ordinary superphosphate is mostly monocalcium phosphate and calcium sulfate. It is made from phosphate rock and sulfuric acid and is equated to a 20% phosphorus pentoxide (P_2O_5) content. The production of normal superphosphate is similar to that for the manufacture of wet-process phosphoric acid except that there is only partial neutralization.

$$CaF_2 \cdot 3Ca_3(PO_4)_2 + 17H_2O + 7H_2SO_4 \rightarrow 3[CaH_4(PO_4)_2 \cdot H_2O]$$
$$+2HF + 7(CaSO_4 \cdot 2H_2O)$$

Triple superphosphate, made from phosphate rock and phosphoric acid, is mostly monocalcium and dicalcium phosphate.

$$CaF_2 \cdot 3Ca_3(PO_4)_2 + 14H_3PO_4 \rightarrow 10CaH_4(PO_4)_2 + 2HF$$

Diammonium phosphate (DAP) is made from wet-process phosphoric acid of about 40% phosphorus pentoxide content and ammonia.

When an ammonia fertilizer is mixed with a superphosphate, a chemical reaction occurs that changes the active ingredient's structure.

$$H_3PO_4 + NH_3 \rightarrow NH_4H_2PO_4$$

$$Ca(H_2PO_4)_2 \cdot H_2O + NH_3 \rightarrow CaHPO_4 + NH_4H_2PO_4 + H_2O$$

$$NH_4H_2PO_4 + NH_3 \rightarrow (NH_4)_2HPO_4$$

$$2CaHPO_4 + CaSO_4 + 2NH_3 \rightarrow Ca(PO_4)_2 + (NH_4)_2SO_4$$

$$NH_4H_2PO_4 + CaSO_4 + NH_3 \rightarrow CaHPO_4 + (NH_4)_2SO_4$$

Slow-release or controlled-release fertilizers make application requirements less stringent. Urea-formaldehyde resins in combination with nitrogen fertilizers tie up the nitrogen for a longer time, whereas degradation of the polymer occurs slowly by sunlight. This type of fertilizer is especially popular for the high nitrogen content of home lawn fertilizers.

FLUORINE

Fluorine (melting point: $-219.6°C$, boiling point: $-188.4°C$) is a pale greenish-yellow reactive gas that occurs in combined form in fluorine-containing minerals such as fluorspar, fluorapatite, and cryolite. Fluorine is also produced by the electrolysis of potassium bifluoride (KHF_2 or KF·HF) under varying conditions of temperature and electrolyte composition.

Fluorspar is also used to manufacture hydrogen fluoride (hydrofluoric acid, aqueous and anhydrous) is manufactured in heated kilns by the reaction of fluorspar with sulfuric acid.

$$CaF_2 + H_2SO_4 \rightarrow CaSO_4 + 2HF$$
fluorspar

The hot, gaseous hydrogen fluoride is either absorbed in water or liquefied; refrigeration is employed to obtain the anhydrous product needed for fluorocarbon manufacture and other uses. Although hydrofluoric acid is corrosive, concentrations of 60% and above can be handled in steel at lower temperatures; lead, carbon, and special alloys are also used in the process equipment.

The largest production of fluorine compounds is that of hydrofluoric acid (anhydrous and aqueous), used in making *alkylate* for gasoline manufacture. It is also employed in the preparation of inorganic fluorides, elemental fluorine, and many organic fluorine- and non-fluorine-containing compounds. Aqueous hydrogen fluoride is used in the glass, metal, and petroleum industries and in the manufacture of many inorganic and acid fluorides. Three of the most unusual plastics known are Teflon, a polymerization product of tetrafluorethylene ($CF_2{=}CF_2$).

Fluorine is also used for the manufacture of sulfur hexafluoride (SF_6) for high-voltage insulation and for uranium hexafluoride. Fluorine is used directly or combined with higher metals (cobalt, silver, cerium, etc.) and halogens (chlorine and bromine) for organic fluorinations and the growing production of fluorocarbons.

Hydrogen fluoride is used to prepare fluorocarbons and one-third of the total goes to the aluminum industry, where synthetic cryolite, sodium aluminum fluoride, is a major constituent of the electrolyte. It is also consumed in the melting and refining of secondary aluminum. Other uses of hydrofluoric acid are found in the metals and petroleum industries.

FLUOROCARBONS

Fluorocarbons are compounds of carbon, fluorine, and chlorine with little or no hydrogen. Fluorocarbons containing two or more fluorine atoms on a carbon atom are characterized by extreme chemical inertness and stability. Their volatility and density are greater than those of the corresponding hydrocarbons. However, environmental regulations have restricted the use of many of these compounds.

Fluorocarbons are made from chlorinated hydrocarbons by reacting them with anhydrous hydrogen fluoride, using an antimony pentachloride ($SbCl_5$) catalyst.

The fluorocarbons trichlorofluoromethane, dichlorodifluoromethane, and chlorodifluoromethane are major fluorocarbon compounds.

$$CCl_4 + HF \rightarrow CCl_3F + HCl$$

$$CCl_4 + 2HF \rightarrow CCl_2F_2 + 2HCl$$

Difluoromonochloromethane is made by substituting chloroform for the carbon tetrachloride.

$$CHCl_3 + 2HF \rightarrow CHClF_2 + 2HCl$$

In the process (Fig. 1), anhydrous hydrogen fluoride and carbon tetrachloride (or chloroform) are bubbled through molten antimony pentachloride catalyst in a steam-jacketed atmospheric pressure reactor at 65 to 95°C. The gaseous mixture of fluorocarbon and unreacted chlorocarbon is distilled to separate and recycle the chlorocarbon to the reaction. Waste hydrogen chloride is recycled by use of water absorption and the last traces of hydrogen chloride and chlorine are removed in a caustic scrubbing tower.

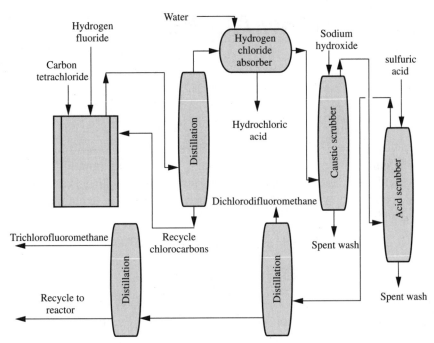

FIGURE 1 Fluorocarbon manufacture.

FORMALDEHYDE

Formaldehyde (methanal, melting point: –92°C, boiling point: –21°C) is produced solely from methanol by using a silver catalyst (Fig. 1) or a metal oxide catalyst (Fig. 2). Either process can be air oxidation or simple dehydrogenation.

$$2CH_3OH + O_2 \rightarrow 2HCH{=}O + 2H_2O$$

$$CH_3OH \rightarrow HCH{=}O + H_2$$

These two reactions occur simultaneously in commercial units in a balanced autothermal reaction because the oxidative reaction furnishes the heat to cause the dehydrogenation to take place.

In the process (Figs. 1 and 2), fresh and recycle methanol are vaporized, superheated, and passed into the methanol-air mixer. Atmospheric air is purified, compressed, and preheated to 54°C in a finned heat exchanger.

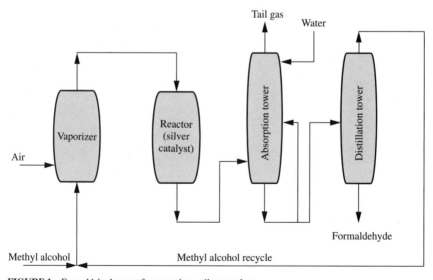

FIGURE 1 Formaldehyde manufacture using a silver catalyst.

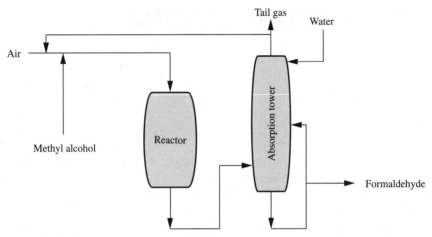

FIGURE 2 Formaldehyde manufacture using a metal oxide catalyst.

The products leave the converter (a water-jacketed vessel containing the catalyst) at 620°C and at 34 to 69 kPa absolute. About 65 percent of the methanol is converted per pass. Temperatures are on the order of 450 to 900°C and there is a short contact time of 0.01 second.

The reactor effluent contains about 25% formaldehyde, which is absorbed with the excess methanol and piped to the make tank. The latter feeds the methanol column for separation of recycle methanol overhead, the bottom stream containing the formaldehyde and a few percent methanol. The water intake adjusts the formaldehyde to 37% strength (marketed as formalin). The catalyst is easily poisoned so stainless-steel equipment must be used to protect the catalyst from metal contamination.

In the pure form, formaldehyde in the pure form is a gas with a boiling point of –21°C but is unstable and readily trimerizes to trioxane or polymerizes to paraformaldehyde. Formaldehyde is stable only in water solution, commonly 37 to 56% formaldehyde by weight and often with methanol (3 to15%) present as a stabilizer.

FUROSEMIDE

Furosemide, 4-chloro-N-furfuryl-5-sulfamoyl anthranilic acid, is prepared by treating 2,4,5-trichlorobenzoic acid with chlorosulfonic acid, and further treatment with ammonia and furfuryl amine.

Furosemide can also be synthesized starting with 2,4-dichlorobenzoic acid (formed by chlorination and oxidation of toluene). Reaction with chlorosulfonic acid is an electrophilic aromatic substitution via the species -SO$_2$Cl⁻ attacking ortho and para to the chlorines and meta to the carboxylate. Ammonolysis to the sulfonamide is followed by nucleophilic aromatic substitution of the less hindered chlorine by furfurylamine (obtained from furfural—a product obtained by the hydrolysis of carbohydrates).

Furosemide is used as a diuretic and blood pressure reducer.

GASOLINE

Gasoline, also called *gas* (United States and Canada) or *petrol* (Great Britain), or *benzine* (Europe), is a mixture of volatile, flammable liquid hydrocarbons derived from petroleum and used as fuel for internal-combustion engines.

The hydrocarbons in gasoline boil below 180°C (355°F) or, at most, below 200°C (390°F). The hydrocarbon constituents in this boiling range are those that have 4 to 12 carbon atoms in their molecular structure and are classified into three general types: paraffins (including the cycloparaffins and branched materials), olefins, and aromatics.

Highly branched paraffins, which are particularly valuable constituents of gasolines, are not usually the principal paraffinic constituents of straight-run gasoline. The more predominant paraffinic constituents are usually the normal (straight-chain) isomers, which may dominate the branched isomers by a factor of 2 or more. This is presumed to indicate the tendency to produce long uninterrupted carbon chains during petroleum maturation rather than those in which branching occurs.

Gasoline is manufactured by distillation in which the volatile, more valuable fractions of crude petroleum are separated. Later processes, known as *cracking*, were designed to raise the yield of gasoline from crude oil by converting the higher-molecular-weight constituents of petroleum into lower-molecular-weight products Other methods used to improve the quality of gasoline and increase its supply include *polymerization, alkylation, isomerization,* and *reforming*.

Polymerization is the conversion of gaseous olefins, such as propylene and butylene, into larger molecules in the gasoline range. A*lkylation* is a process combining an olefin and a paraffin such as *iso*-butane). *Isomerization* is the conversion of straight-chain hydrocarbons to branched-chain hydrocarbons. *Reforming* is the use of either heat or a catalyst to rearrange the molecular structure.

Aviation gasoline, now usually found in use in light aircraft and older civil aircraft, has a narrower boiling range than conventional (automobile) gasoline, that is, 38 to 170°C (100 to 340°F) compared to approximately

−1 to 200°C (30 to 390°F) for automobile gasoline. The narrower boiling range ensures better distribution of the vaporized fuel through the more complicated induction systems of aircraft engines. Aircraft operate at altitudes at which the prevailing pressure is less than the pressure at the surface of the earth (pressure at 17,500 feet is 7.5 psi compared to 14.7 psi at the surface of the earth). Thus, the vapor pressure of aviation gasoline must be limited to reduce boiling in the tanks, fuel lines, and carburetors. Thus, the aviation gasoline does not usually contain the gaseous hydrocarbons (butanes) that give automobile gasoline the higher vapor pressures.

Methanol and a number of other alcohols and ethers are considered high-octane enhancers of gasoline. They can be produced from various hydrocarbon sources other than petroleum and may also offer environmental advantages insofar as the use of oxygenates would presumably suppress the release of vehicle pollutants into the air.

Of all the oxygenates, methyl-*t*-butyl ether (MTBE) is attractive for a variety of technical reasons. It has a low vapor pressure, can be blended with other fuels without phase separation, and has the desirable octane characteristics. If oxygenates achieve recognition as vehicle fuels, the biggest contributor will probably be methanol, the production of which is mostly from synthesis gas derived from methane.

Other additives to gasoline often include detergents to reduce the buildup of engine deposits, anti-icing agents to prevent stalling caused by carburetor icing, and antioxidants (oxidation inhibitors) used to reduce *gum* formation.

GLASS

Glass is a rigid, undercooled liquid having no definite melting point and a sufficiently high viscosity to prevent crystallization that results from the union of the nonvolatile inorganic oxides, sand, and other constituents, and thus is a product with *random* atomic structure.

In order to produce the various glasses, soda ash, salt cake, and limestone or lime are required to flux the silica. In addition, there is a contribution of lead oxide, pearl (as potassium carbonate), saltpeter, borax, boric acid, arsenic trioxide, feldspar, and fluorspar, together with a great variety of metallic oxides, carbonates, and the other salts required for colored glass.

GLUTAMIC ACID

See Monosodium Glutamate.

GLYCEROL

Glycerol (glycerin, melting point: 18°C, boiling point: 290°C, density: 1.2620, flash point: 177°C) is a clear, nearly colorless liquid having a sweet taste but no odor.

Glycerol may be produced by a number of different methods, such as:

1. The saponification of glycerides (oils and fats) to produce soap.
2. The recovery of glycerin from the hydrolysis, or splitting, of fats and oils to produce fatty acids.
3. The chlorination and hydrolysis of propylene and other reactions from petrochemical hydrocarbons.

Natural glycerol is produced as a coproduct of the direct hydrolysis of triglycerides from natural fats and oils in large continuous reactors at elevated temperatures and pressures with a catalyst (Fig. 1). Water flows countercurrent to the fatty acid and extracts glycerol from the fatty phase. The sweet water from the hydrolyzer column contains about 12% glycerol. Evaporation of the sweet water from the hydrolyzer is a much easier operation than with evaporation of spent soap lye glycerin in the kettle process. The high salt content of soap lye glycerin requires frequent soap removal from the evaporators. Hydrolyzer glycerin contains practically no salt and is readily concentrated.

The sweet water is fed to a triple-effect evaporator where the concentration is increased from 12% to 75 to 80% glycerol. After concentration of the sweet water to hydrolyzer crude, the crude is settled for 48 hours at elevated temperatures to reduce fatty impurities that could interfere with subsequent processing. Settled hydrolyzer crude contains approximately 78% glycerol and 22% water. The settled crude is distilled under vacuum at approximately 200°C. A small amount of caustic is usually added to the still feed to saponify fatty impurities and reduce the possibility of codistillation with the glycerol. The distilled glycerin is condensed in three stages at decreasing temperatures. The first stage yields the purest glycerin, usually 99% glycerol and lower-quality grades of glycerin are collected in the

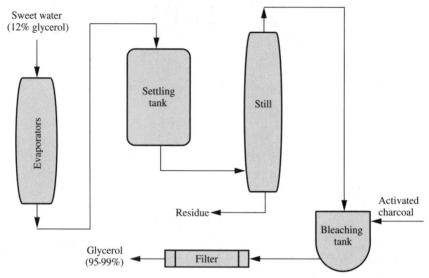

Sweet water
(12% glycerol)

Evaporators

Settling tank

Still

Residue

Activated charcoal

Bleaching tank

Glycerol
(95-99%)

Filter

FIGURE 1 Glycerol manufacture using the sweet water process.

second and third condensers. Final purification of glycerin is accomplished by carbon bleaching, followed by filtration or ion exchange.

There are several synthetic methods for the manufacture of glycerol. One process (Fig. 2) involves chlorination of propylene at 510°C (950°F) to produce allyl chloride in seconds in amounts greater than 85 percent of theory (based on the propylene). Vinyl chloride, some disubstituted olefins, and some 1,2 and 1,3-dichloropropanes are also formed. Treatment of the allyl chloride with hypochlorous acid at 38°C (100°F) produces glycerin dichlorohydrin ($CH_2ClCHClCH_2OH$), which can be hydrolyzed by caustic soda in a 6% Na_2CO_3 solution at 96°C. The glycerin dichlorohydrin can be hydrolyzed directly to glycerin, but this takes two molecules of caustic soda; hence a more economical procedure is to react with the cheaper calcium hydroxide, taking off the epichlorohydrin as an overhead in a stripping column. The epichlorohydrin is easily hydrated to monochlorohydrin and then hydrated to glycerin with caustic soda.

$$CH_3CH=CH_2 + Cl_2 \rightarrow CH_2ClCH=CH_2 + HCl$$

$$CH_2ClCH=CH_2 + HOCl \rightarrow CH_2ClCHClCH_2OH$$

$$CH_2ClCHClCH_2OH + 2NaOH \rightarrow CH_2OHCHOHCH_2OH + 2NaCl$$

The overall yield of glycerin from allyl chloride is above 90 percent.

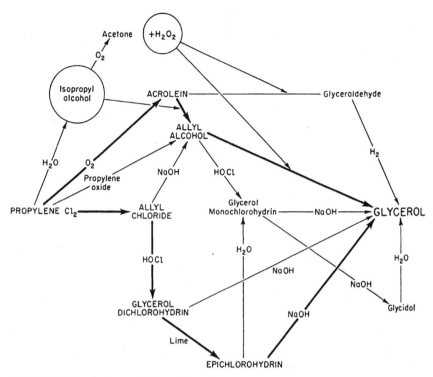

FIGURE 2   Routes for the manufacture of glycerol.

Another process for obtaining glycerol from propylene involves the following reactions, where isopropyl alcohol and propylene furnish acetone and glycerin (through acrolein) in good yield (Fig. 2).

$$CH_3CHOHCH_3 + air \rightarrow CH_3COCH_3 + H_2O$$

$$CH_3CH{=}CH_2 + air \rightarrow CH_2{=}CHCHO + H_2O$$

$$CH_2{=}CHCHO + H_2O_2 \rightarrow CHOCHOHCH_2OH$$

$$CHOCHOHCH_2OH \rightarrow CH_2OHCHOHCH_2OH$$

GRAPHITE

See **Carbon.**

GYPSUM

See Calcium Sulfate.

HELIUM

See **Rare Gases.**

HERBICIDES

Herbicides are a class of compounds that allow chemical methods of weed control. This commenced with the introduction of 2,4-dichlorophenoxyacetic acid (2,4-D) in the mid-1940s.

Phenol is the starting material for 2,4-dichlorophenoxyacetic acid via electrophilic aromatic substitution. Chlorination of phenol gives 2,4-dichlorophenol and the sodium salt of this compound is reacted with sodium chloroacetate and acidification gives 2,4-dichlorophenoxyacetic acid.

Another herbicide, 2,4,5-trichlorophenoxyacetic acid, is synthesized by starting with the chlorination of benzene to give 1,2,4,5-tetrachlorobenzene, which reacts with caustic to give 2,4,5-trichlorophenol. Conversion to the sodium salt followed by reaction with sodium chloroacetate and acidification gives 2,4,5-trichlorophenoxyacetic acid. Agent Orange is a 1-to-1 mixture of the butyl esters of 2,4,5-trichlorophenoxyacetic acid and 2,4,-dichlorophenoxyacetic acid.

The bipyridyl herbicide Paraquat is made by reduction of pyridine to radical ions, which couple at the para positions. Oxidation and reaction with methyl bromide gives paraquat. Diquat is formed by dehydrogenation of pyridine and quaternization with ethylene dibromide.

HEXAMETHYLENEDIAMINE

Hexamethylenediamine (HMDA, boiling point: 204°C, melting point: 41°C) is used in the synthesis of nylon, and it is manufactured from butadiene.

In the process, butadiene first adds one mole of hydrogen cyanide at 60°C with a nickel catalyst via both 1,2 and 1,4 addition to give, respectively, 2-methyl-3-butenonitrile and 3-pentenonitrile in a 1:2 ratio. Isomerization of the 2-methyl-3-butenonitrile to 3-pentenonitrile takes place at 150°C. Then more hydrogen cyanide, more catalyst, and a triphenylboron promotor react with 3-pentenonitrile to form methylglutaronitrile and mostly adiponitrile. The adiponitrile is formed from 3-pentenonitrile probably through isomerization of 3-pentenonitrile to 4-pentenonitrile and followed by addition of hydrogen cyanide.

$$CH_2=CHCH=CH_2 + 2HCN \rightarrow NC(CH_2)_4CN$$

Extraction and distillation is necessary to obtain pure adiponitrile. Even then the hexamethylenediamine made by hydrogenation of adiponitrile must also be distilled through seven columns to purify it before polymerization to nylon. Hexamethylenediamine is produced from adiponitrile by hydrogenation.

$$NC(CH_2)_4CN + H_2 \rightarrow H_2N(CH_2)_6NH_2$$

Evaporation of the reaction product of formaldehyde and ammonia also produces hexamethylenetetramine.

Hexamethylenediamine is used in the production of nylon 6,6 and mainly in making phenol-formaldehyde resins, where it is known as *hexa*. It is also used as a urinary antiseptic (Urotropine) as well as in the rubber industry and for the manufacture of the explosive cyclonite.

HEXAMETHYLENETETRAMINE

*See **Hexamine**.*

HEXAMINE

Hexamine [hexamethylenetetramine, methenamine, and urotropine, $(CH_2)_6N_4$; melting point: 280°C] is a white crystalline solid that decomposes at higher temperatures. Hexamine is soluble in water but only very slightly soluble in alcohol or ether.

Hexamine is manufactured from anhydrous ammonia (NH_3) and a 45% solution of methanol-free formaldehyde (HCH=O). These raw materials, plus recycle mother liquor, are charged continuously at carefully controlled rates to a high-velocity reactor, since the reaction is exothermic. The reactor effluent is discharged into a vacuum evaporator that also serves to crystallize the product, and the hexamine crystals are washed, dried, and screened. Typically, the yield of hexamine is on the order of 96%.

Although used to some extent in medicine as an internal antiseptic, the primary use of hexamine is in the manufacture of synthetic resins where the compound is a substitute for formalin (aqueous solution of paraformaldehyde) and its sodium hydroxide catalyst. Hexamine is also used as an accelerator for rubber.

HEXANES

Hexane isomers (C_6H_{14}) are produced by two-tower distillation of straight-run gasoline that has been distilled from crude oil or natural gas liquids.

Hexanes are mostly used in gasoline. They are also used as a solvent and as a medium for various polymerization reactions.

HEXYLRESORCINOL

Hexylresorcinol (1,3-dihydroxy-4-hexylbenzene) is an odorless solid that has marked germidical properties and is used as an antiseptic, commonly employed in a dilution of 1:1000.

In the manufacture of hexylresorcinol, resorcinol and caproic acid are heated with a condensing agent, such as zinc chloride, and the intermediate ketone derivative is formed, which is purified by vacuum distillation. After reduction with zinc amalgam and hydrochloric acid, impure hexylresorcinol is formed, which can be purified by vacuum distillation.

HYDROCHLORIC ACID

Hydrogen chloride (HCl, boiling point: –35°C) is a colorless, poisonous gas with a pungent odor. Aqueous solutions of hydrogen chloride are known as *hydrochloric acid* or, if the hydrogen chloride in solution is of the commercial grade, as *muriatic acid*. Hydrochloric acid typically contains 24 to 36% by weight hydrogen chloride.

Hydrochloric acid is obtained from four major sources:

1. As a by-product in the chlorination of both aromatic and aliphatic hydrocarbons or from the thermal degradation of organic chlorine compounds,

$$CH_4 + Cl_2 \rightarrow CH_3Cl + HCl$$

$$CH_2ClCH_2Cl \rightarrow CH_2=CHCl + HCl$$

2. From reacting sodium chloride (salt) and sulfuric acid,

$$2NaCl + H_2SO_4 \rightarrow Na_2SO_4 + 2HCl$$

3. From the combustion of hydrogen and chlorine,

$$Cl_2 + H_2 \rightarrow 2HCl$$

4. From Hargreaves-type operations,

$$4NaCl + 2SO_2 + O_2 + 2H_2O \rightarrow 2Na_2SO_4 + 4HCl$$

The reaction between hydrogen and chlorine is highly exothermic and spontaneously goes to completion as soon as it is initiated. The equilibrium mixture contains about 4% by volume free chlorine. As the gases are cooled, the free chlorine and free hydrogen combine rapidly so that when 200°C is reached, the gas is almost pure hydrogen chloride. By carefully controlling the operating conditions, a gas containing 99% hydrogen chloride can be produced and it can be further purified by absorbing it in water in a tantalum or impervious or impregnated graphite absorber. The aqueous solution is stripped of hydrogen chloride under slight pressure, giving strong gaseous hydrogen chloride that is dehydrated to 99.5% hydrogen chloride by cooling it to $-12°C$. Large amounts of anhydrous hydrogen chloride are needed for preparing methyl chloride, ethyl chloride, vinyl chloride, and other such compounds.

Hydrochloric acid is replacing sulfuric acid in some applications such as metal pickling, which is the cleaning of metal surfaces by acid etching. It leaves a cleaner surface than sulfuric acid, reacts more slowly, and can be recycled more easily. It is used in chemical manufacture especially for phenol and certain dyes and plastics. In oil well drilling, it increases the permeability of limestone by acidifying the drilling process.

HYDROFLUORIC ACID

Hydrofluoric acid (melting point: –83.1°C, boiling point: 19.5°C) is produced by treating fluorspar (CaF_2) with 20% oleum and heating it with sulfuric acid in a horizontal rotating drum.

$$CaF_2 + H_2SO_4 \rightarrow 2HF + CaSO_4$$

Hydrofluoric acid is used for manufacture of fluorocarbons, including fluoropolymers, chlorofluorocarbons; chemical intermediates including fluoroborates, surfactants, herbicides, and electronic chemicals; aqueous hydrofluoric acid; petroleum alkylation; and uranium processing.

See Fluorine.

HYDROGEN

Hydrogen (boiling point: $-252.8°C$) is primarily manufactured by steam-reforming natural gas (CH_4) or hydrocarbons (C_nH_{2n+2}).

A variety of low-molecular-weight hydrocarbons can be used as feedstock in the steam-reforming process. The reaction occurs in two separate steps: reforming and shift conversion.

Methane:

$$CH_4 + H_2O \rightarrow CO + 3H_2 \quad \text{(reforming)}$$

$$CO + H_2O \rightarrow CO_2 + H_2 \quad \text{(shift conversion)}$$

Propane:

$$C_3H_8 + 3O_2 \rightarrow 3CO_2 + 7H_2 \quad \text{(reforming)}$$

$$3CO + 3H_2O \rightarrow 3CO_2 + 3H_2 \quad \text{(shift conversion)}$$

The reforming step makes a hydrogen–carbon monoxide mixture (*synthesis gas*) that is used to produce a variety of other chemicals.

In the steam-reforming process (Fig. 1), the hydrocarbon feedstock is first desulfurized by heating to 370°C in the presence of a metallic oxide catalyst that converts the organosulfur compounds to hydrogen sulfide. Elemental sulfur can also be removed with activated carbon absorption. A caustic soda scrubber removes the hydrogen sulfide by salt formation in the basic aqueous solution.

$$H_2S + 2NaO \rightarrow Na_2S + 2H_2O$$

Steam is added and the mixture is heated in the furnace at 760 to 980°C and 600 psi over a nickel catalyst. When higher-molecular-weight hydrocarbons are the feedstock, potassium oxide is used along with nickel to avoid larger amounts of carbon formation.

There are primary and secondary furnaces in some plants. Air can be added to the secondary reformers. Oxygen reacts with some of the hydrocarbon feedstock to keep the temperature high. The nitrogen of the air is

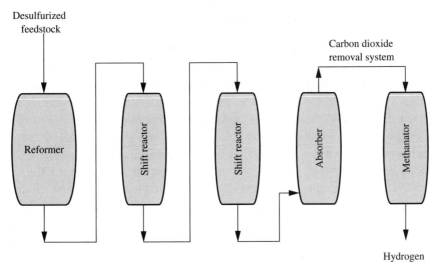

FIGURE 1 Hydrogen production by steam reforming hydrocarbon feedstocks.

utilized when it, along with the hydrogen formed, reacts in the ammonia synthesizer. More steam is added and the mixture enters the shift converter, where iron or chromic oxide catalysts at 425°C further react the gas to hydrogen and carbon dioxide.

Some shift converters have high- and low-temperature sections, the high-temperature section converting most of the carbon monoxide to carbon dioxide. Cooling to 38°C is followed by carbon dioxide absorption with monoethanolamine ($HOCH_2CH_2NH_2$). The carbon dioxide (an important by-product) is desorbed by heating the monoethanolamine and reversing this reaction.

$$HOCH_2CH_2NH_2 + CO_2 + H_2O \rightarrow HOCH_2CH_2NH_3^+HCO_3^-$$

Alternatively, hot carbonate solutions can replace the monoethanolamine. A methanator converts the last traces of carbon dioxide to methane, a less interfering contaminant in hydrogen used for ammonia manufacture.

Hydrogen is also produced by an electrolytic process that produces high-purity hydrogen and consists of passing direct current through an aqueous solution of alkali, and decomposing the water.

$$2H_2O \rightarrow 2H_2 + O_2$$

A typical commercial cell electrolyzes a 15% sodium hydroxide (NaOH) solution, uses an iron cathode and a nickel-plated-iron anode, has an asbestos diaphragm separating the electrode compartments, and operates at temperatures from 60 to 70°C. The nickel plating of the anode reduces the oxygen overvoltage.

Partial oxidation processes rank next to steam-hydrocarbon processes in the amount of hydrogen made. They can use natural gas, refinery gas, or other hydrocarbon gas mixtures as feedstocks, but their chief advantage is that they can also accept liquid hydrocarbon feedstocks such as gas oil, diesel oil, and even heavy fuel oil. All processes employ noncatalytic partial combustion of the hydrocarbon feed with oxygen in the presence of steam in a combustion chamber at flame temperatures between 1300 and 1500°C. For example, with methane as the principal component of the feedstock:

$$CH_4 + 2O_2 \rightarrow CO_2 + 2H_2O$$

$$CH_4 + CO_2 \rightarrow 2CO + 2H_2$$

$$CH_4 + H_2O \rightarrow CO + 3H_2$$

The overall process is a net producer of heat; for efficient operation, heat recovery (using waste heat boilers) is important.

Most of the hydrogen is generated on site for use by various industries, particularly the petroleum industry. Other uses include ammonia production, metallurgical industries to reduce the oxides of metals to the free metals, methanol production, and hydrogen chloride manufacture.

HYDROGEN CYANIDE

Hydrogen cyanide (melting point: $-14°C$, boiling point: $26°C$) is manufactured by the reaction of natural gas (methane), ammonia, and air over a platinum or platinum-rhodium catalyst at elevated temperature (the Andrussow process).

$$2CH_4 + 2NH_3 + O_2 \rightarrow 2HCN + 6H_2O$$

Hydrogen cyanide is also available as a by-product from acrylonitrile manufacture by ammoxidation.

$$CH_2=CHCH_3 + 2NH_3 + 3O_2 \rightarrow 2CH_2=CH-CN + 6H_2O \ (+ HCN)$$

Hydrogen cyanide is used for the production of methyl methacrylate, adiponitrile, cyanuric chloride, and chelating agents.

*See **Methane**.*

HYDROGEN PEROXIDE

Hydrogen peroxide is the most widely used peroxide compound. Originally, it was produced by the reaction of barium peroxide and sulfuric acid but this process and use have been superseded.

The most important method of making hydrogen peroxide is by reduction of anthraquinone to the hydroquinone, followed by reoxidation to anthraquinone by oxygen and formation of the peroxide.

The hydrogen peroxide is extracted with water and concentrated, and the quinone is recycled for reconversion to the hydroquinone. A second organic process uses isopropyl alcohol, which is oxidized at moderate temperatures and pressures to hydrogen peroxide and acetone. After distillation of the acetone and unreacted alcohol, the residual hydrogen peroxide is concentrated.

Hydrogen peroxide applications include commercial bleaching dye oxidation, the manufacture of organic and peroxide chemicals. Hydrogen peroxide is also used in pulp and paper chemical synthesis, textiles, and environmental control, including municipal and industrial water treatment.

IBUPROFEN

Ibuprofen, which is sold under trade names such as Motrin® and Advil®, is an alternative to aspirin and acetaminophen because of its analgesic and antiinflammatory properties.

Ibuprofen can be synthesized from isobutylbenzene by a Friedel-Crafts acylation with acetyl chloride, followed by formation of a cyanohydrin. Treatment with hydrogen iodide and phosphorus reduces the benzylic hydroxyl to a hydrogen and hydrolyzes the nitrile to a carboxylic acid.

INSECTICIDES

Insecticides are chemical compounds that are used for the control of insects either through death of the insect or through interference with the reproductive cycle of the insect.

Early insecticides also included organic natural products such as nicotine, rotenone, and pyrethrin. Rotenone is used as a method of killing rough fish when a lake has been taken over completely by them. In a couple of weeks after treatment the lake is then planted with fresh game fish. The pyrethrins, originally obtained from Asian or Kenyan flowers, can now also be synthesized. Nicotine is no longer used as an insecticide because it is not safe for humans.

Dichlorodiphenyltrichloroethane (DDT) is no longer being used in large amounts because of its persistence in the environment, although for many uses there were no good substitutes available. DDT was first made in 1874 but its insecticidal properties were not discovered until 1939.

Second-generation insecticides are of three major types: chlorinated hydrocarbons, organophosphorus compounds, and carbamates. Synthetic pyrethroids are a recent fourth type. A very dramatic decline of the chlorinated hydrocarbons in the late 1960s and the 1970s, while the use of organophosphates and carbamates increased.

The use of chlorinated hydrocarbons has declined worldwide and is banned in many countries for three main reasons: (1) concern over the buildup of residues, (2) the increasing tendency of some insects to develop resistance to the materials, and (3) the advent of insecticides that can replace the organochlorine compounds.

In the 1970s, organophosphorus compounds became the leading type of insecticide. Over 40 such compounds have been registered in the United States as insecticides. The first organophosphorus insecticide was synthesized in 1938 and is known as tetraethyl pyrophosphate (TEPP). Another phosphate insecticide, Malathion is synthesized by condensing diethyl maleate with the o,o-dimethyl phosphorodithioic acid.

The 1950s saw the development of carbaryl (Sevin®), the first major carbamate and is manufactured by condensing l-naphthol with methyl isocyanate. The l-naphthol is made from naphthalene by hydrogenation, oxidation, and dehydrogenation. The 1-naphthol is made from naphthalene, which is obtained from coal tar distillation or from petroleum.

Methyl isocyanate can be made from phosgene ($COCl_2$) and methylamine (CH_3NH_2), which would circumvent use of the isocyanate. Methyl isocyanate is a very dangerous chemical and was responsible for the deaths of over 2500 people in the worst industrial accident ever, that of the carbamate insecticide plant in Bhopal, India on December 3, 1984.

INSULIN

Insulin, a hormone, plays a key role in catalyzing the processes by which glucose (carbohydrates) furnishes energy or is stored in the body as glycogen or fat. The absence of insulin not only interrupts these processes, but produces depression of essential functions and, in extreme cases, even death. Insulin protein is characterized by a high sulfur content in the form of cystine and it is unstable in alkaline solution.

Insulin is isolated from the pancreas of beef or pigs and by extraction with acidified alcohol, followed by purification. In the process (Fig. 1), the crude alcoholic extract is run from two strong extraction-centrifuge units into a collection tank from which the extract is neutralized with ammonia and filter aid added. In a continuous precoat drum filter, the cake is separated and washed, the clear liquor going to the reacidification tank. In evaporators, the first stage removes alcohol, with subsequent waste-fat separation. The extract goes to a chill tank, with filter aid added, through a filter press and into the second evaporator. From the second evaporator, the concentrated extract is filtered and conducted to the first salting-out tank, followed by filter-press filtration with filtrate to sewer and salt cake to purification for the second salting out. The second-salting-out product is crystallized twice to furnish Iletin® (insulin) crystals.

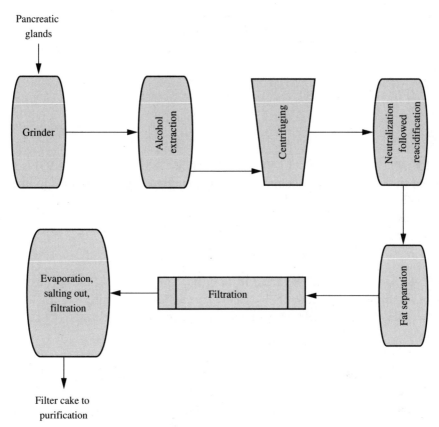

FIGURE 1   Manufacture of insulin.

IODINE

Iodine (melting point: 113.5°C, boiling point: 184.4°C, density: 4.93) is a red to purple solid that sublimes readily under ambient conditions. Iodine can be produced from iodates (Fig. 1):

$$IO_3^- + 3SO_2 + 3H_2O \rightarrow I^- + 3SO_4^{2-} + 6H^+$$

$$IO_3^- + I^- + 6H^+ \rightarrow 3I_2 + 3H_2O$$

Iodine can also be produced from brine. This process (Fig. 2) consists of cleaning the solution (of clays and other materials), adding sulfuric acid to a pH <2.5 followed by treatment with gaseous chlorine:

$$2I^- + Cl_2 \rightarrow I_2 + 2Cl^-$$

after which the iodine is recovered by a countercurrent air blow out step. process.

Iodine is used for the manufacture of organic compounds, for the manufacture of potassium iodide and sodium iodide, and for the manufacture of other inorganic compounds. Iodine is used as a catalyst in the chlorination of organic compounds and in analytical chemistry for determination of the *iodine numbers* of oils. Iodine for medicinal, photographic, and pharmaceutical purposes is usually in the form of alkali iodides, prepared through the agency of ferrous iodide.

In addition, iodine is also used for the manufacture of dyes and as a germicide. Simple iodine derivatives of hydrocarbons, such as iodoform (CHI_3), have an antiseptic action. Organic compounds containing iodine have been used as rubber emulsifiers, chemical antioxidants, and dyes and pigments.

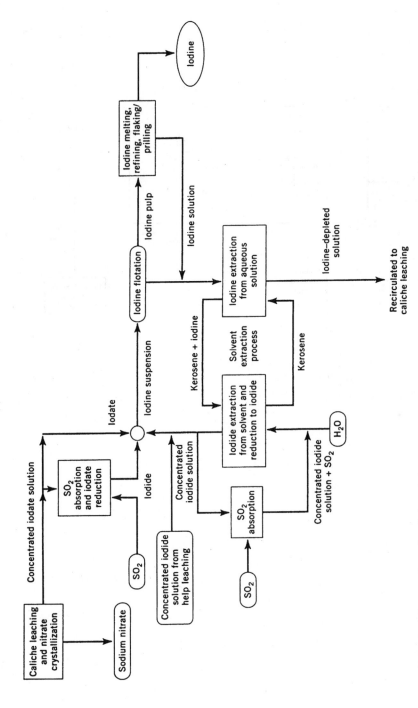

FIGURE 1 Iodine manufacture from iodate solutions.

2.277

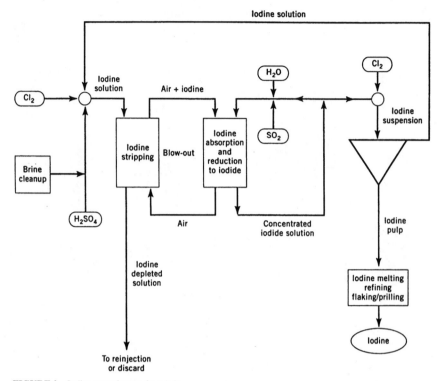

FIGURE 2   Iodine manufacture from brine.

ISONIAZID

Isoniazid, isonicotinic acid hydrazide, is the most potent and selective of the tuberculostatic antibacterial agents.

ISOPRENE

Isoprene (melting point: −146°C, boiling point: 34°C, density: 0.6810) may be produced by the dehydrogenation of *iso*-pentane in the same plant used for the production of butadiene. However, the presence of 1,3-pentadiene (for which there is very little market) requires a purification step. One method produces isoprene from propylene. Thus, dimerization of propylene to 2-methyl-1-pentene is followed by isomerization of the 2-methyl-1-pentene to 2-methyl-2-pentene, which upon pyrolysis gives isoprene and methane.

$$CH_3CH=CH_2 \rightarrow CH_3CH_2CH_2CH(CH_3)=CH_2$$

$$CH_3CH_2CH_2CH(CH_3)=CH_2 \rightarrow CH_3CH_2CH=C(CH_3)_2$$

$$CH_3CH_2CH=C(CH_3)_2 \rightarrow CH_2=CHC(CH_3)=CH_2 + CH_4$$

Isoprene can be also produced from isobutylene and formaldehyde and the product is of exceptional purity when made by this method. The *iso*-butylene is first condensed with formaldehyde to yield the cyclic 4,4-dimethyl-*m*-dioxane, which produces isoprene.

$$(CH_3)_2C=CH_2 + 2HCH=O \rightarrow (CH_3)_2COCH_2OCH_2CH_2$$

$$(CH_3)_2COCH_2OCH_2CH_2 \rightarrow CH_2=CHC(CH_3)=CH_2 + HCH=O + H_2O$$

The production of isoprene from acetylene and acetone is also possible. The acetylene is reacted with acetone to produce 2-methyl-3-butyn-2-ol which, by hydrogenation produces 2-methyl-3-butene-2-ol. Dehydration then yields isoprene.

$$HC\equiv CH + (CH_3)_2C=O \rightarrow (CH_3)_2C(OH)C\equiv CH$$

$$(CH_3)_2C(OH)C\equiv CH + H_2 \rightarrow (CH_3)_2C(OH)CH=CH_2$$

$$(CH_3)_2C(OH)CH=CH_2 \rightarrow CH_2=CHC(CH_3)=CH_2 + H_2O$$

ISO-PROPYL ALCOHOL

Iso-propyl alcohol (2-propanol, *iso*-propanol, rubbing alcohol) is manufactured by the esterification/hydrolysis of propylene to *Iso*-propyl alcohol. Unlike ethanol, for which the esterification/hydrolysis has been replaced by direct hydration, the direct process for *Iso*-propyl alcohol is more difficult for crude propylene.

In the esterification process only the propylene reacts and conditions can be maintained so that ethylene is inert.

$$CH_3CH=CH_2 + H_2SO_4 \rightarrow CH_3CH(OSO_3H)CH_3$$

$$CH_3CH(OSO_3H)CH_3 + H2O \rightarrow CH_3CH(OH)CH_3 + H_2SO_4$$

The esterification step occurs with 85% sulfuric acid at 24 to 27°C, and dilution to 20% concentration is done in a separate tank. The *iso*-propyl alcohol is distilled from the dilute acid that is concentrated and returned to the esterification reactor. The *Iso*-propyl alcohol is originally distilled as a 91% azeotrope with water. Absolute *iso*-propyl alcohol, boiling point 82.5°C, is obtained by distilling a tertiary azeotrope with isopropyl ether. A 95% yield is realized.

Iso-propyl alcohol is used to produce acetone, pharmaceuticals, processing solvents, and coatings. Some of the chemicals derived from *iso*-propyl alcohol are *iso*-propyl ether (an industrial extraction solvent), *iso*-propyl acetate (a solvent for cellulose derivatives), *iso*-propyl myristate (an emollient, lubricant, and blending agent in cosmetics, inks, and plasticizers), *t*-butylperoxy *iso*-propyl carbonate (a polymerization catalyst and curing agent), and *iso*-propylamine and *diiso*-propylamine (low-boiling bases).

ISOQUINOLINE

*See **Quinoline.***

KEROSENE

Kerosene (*kerosine, paraffin oil*; approximately boiling range: 205 to 260°C, flash point: approximately 25°C) is a flammable pale-yellow or colorless oily liquid with a characteristic odor. The term *kerosene* is also too often incorrectly applied to various fuel oils, but a fuel oil is actually any liquid or liquid petroleum product that produces heat when burned in a suitable container or that produces power when burned in an engine.

Kerosene is a mixture of hydrocarbons (>C_{12} and higher) that was first manufactured in the 1850s from coal tar, hence the name coal oil is often applied to kerosene, but petroleum became the major source after 1859. From that time, kerosene fraction has remained a product of petroleum. However, the quantity and quality vary with the type of crude oil, and although some crude oils yield excellent kerosene quite simply, others produce kerosene that requires substantial refining.

Kerosene is now produced from petroleum either by distillation or by cracking the less volatile portion of crude oil at atmospheric pressure and elevated temperatures.

Kerosene is used for burning in lamps and domestic heaters or furnaces, as a fuel or fuel component for jet engines, and as a solvent for greases and insecticides.

KEVLAR

See Polyamides.

KRYPTON

*See **Rare Gases.***

LACTIC ACID

Lactic acid (2-hydroxypropionic acid, $CH_3CHOHCO_2H$, boiling point: 122°C, melting point: 18°C, density: 1.2060) is one of the oldest known organic acids. It is the primary acid constituent of sour milk and is formed by the fermentation of milk sugar (lactose) by *Streptococcus lactis*.

Commercially, lactic acid is manufactured by controlled fermentation of the hexose sugars from molasses, corn, or milk. Lactates are made by synthetic methods from acetaldehyde and lactonitrile, a by-product acrylonitrile production.

LEAD AZIDE

See *Explosives.*

LEAD CARBONATE

Lead carbonate ($PbCO_3$) forms colorless orthorhombic crystals and it decomposes at about 315°C. It is nearly insoluble in cold water, but is transformed in hot water to the basic carbonate, $2PbCO_3 \cdot Pb(OH)_2$. Lead carbonate is soluble in acid and alkali, but insoluble in alcohol and ammonia.

Lead carbonate is prepared by treating an aqueous slurry of lead oxide with acetic acid in the presence of air and carbon dioxide:

$$PbO + CO_2 \rightarrow PbCO_3$$

or by shaking a suspension of a lead salt less soluble than the carbonate with ammonium carbonate at a low temperature to avoid formation of basic lead carbonate.

Basic lead carbonate (white lead, $2PbCO_3 \cdot Pb(OH)_2$) forms white hexagonal crystals; it decomposes when heated to 400°C. Basic lead carbonate is insoluble in water and alcohol, slightly soluble in carbonated water, and soluble in nitric acid.

It is produced by several methods, in which soluble lead acetate is treated with carbon dioxide. For example, treatment of an aqueous slurry of finely divided lead metal or monoxide, or a mixture of both, with acetic acid in the presence of air and carbon dioxide produces very fine particle-size basic lead carbonate and ranges in carbonate content from 62 to 65% (theoretical: 68.9% $PbCO_3$).

Lead carbonate has a wide range of applications. It catalyzes the polymerization of formaldehyde to high molecular weight crystalline poly(oxymethylene) products. It is used in poly(vinyl chloride) friction liners for pulleys on drive cables of hoisting engines. To improve the bond of polychloroprene to metals in wire-reinforced hoses, 10 to 25 parts of lead carbonate are used in the elastomer. Lead carbonate is used as a component of high-pressure lubricating greases, as a catalyst in the curing of moldable thermosetting silicone resins, as a coating on vinyl chloride polymers to improve their dielectric properties, as a component of corrosion-resistant, dispersion-strengthened grids in lead-acid storage batteries, as a photoconductor for electrophotography, as a coating on heat-sensitive sheets for thermographic copying, as a component of a lubricant-stabilizer for poly(vinyl chloride), as a component in the manufacture of thermistors, and as a component in slip-preventing waxes for steel cables to provide higher wear resistance.

Basic lead carbonate has many other uses, including as a catalyst for the preparation of polyesters from terephthalic acid and diols, a ceramic glaze component, a curing agent with peroxides to form improved polyethylene wire insulation, a pearlescent pigment, color-changing component of temperature-sensitive inks, a red-reflecting pigment in iridescent plastic sheets, a smudge-resistant film on electrically sensitive recording sheets, a lubricating grease component, a component of ultraviolet light reflective paints to increase solar reflectivity, an improved cool gun-propellant stabilizer that decomposes and forms a lubricating lead deposit, a heat stabilizer for poly(vinyl chloride) polymers, and a component of weighted nylon-reinforced fish nets made of poly(vinyl chloride) fibers.

LEAD CHROMATE

Basic lead chromate [PbCrO$_4$·Pb(OH)$_2$] may be used as an orange-red pigment; it is an excellent corrosion inhibitor. It is manufactured by boiling white lead (PbCO$_3$) with a solution of sodium dichromate.

LEAD STYPHNATE

See **Explosives.**

LIGNIN

Lignin occurs as a large percentage of the noncellulosic part of wood. Newer laboratory processes yield quality lignin with molecular weights of 200 to 1000, but kraft process lignin has a molecular weight of 1000 to 50,000 and is altered chemically by sulfonation.

Presently it is mostly used as a fuel, but as petroleum becomes increasingly scarce and expensive, proposals and experimental plants for using this material begin to appear. One such process uses fluid-bed hydrocracking and dealkylation to produce phenols and benzene.

Lignin derivatives, sulfonated alkali lignin and sulfite lignosulfonates, are being used to increase tertiary oil recovery in *pumped out* oil wells replacing more expensive synthetic detergents.

Lignosulfonates are metal salts of the sulfonated products of lignin and are a by-product of the pulp and paper industry. Pulping of wood with a sulfite solution dissolves the lignin portion, leaving behind the cellulose fibers that are processed into paper. The sulfite solution is concentrated and the resulting solids sold as lignosulfonates. Only a small amount of the spent sulfite liquor solids is used each year. The rest is burned to recover heat. Liquor from an alkaline sulfate pulping process can be concentrated to give a solid called alkali lignin, but little of this is used except as fuel.

See **Lignosulfonates.**

LIGNOSULFONATES

Lignosulfonates are metal salts of the sulfonated products of lignin and are a by-product of the pulp and paper industry. Pulping of wood with a sulfite solution dissolves the lignin portion, leaving behind the cellulose fibers that are processed into paper. The sulfite solution is concentrated and the resulting solids sold as lignosulfonates. Only a small amount of the spent sulfite liquor solids is used each year. The rest is burned to recover heat. A molecular weight of 250 for the lignosulfonate monomer is approximate, but the product may contain material with a molecular weight as high as 100,000. Liquor from an alkaline sulfate pulping process can be concentrated to give a solid called alkali lignin, but little of this is used except as fuel.

The uses of lignosulfonates include the manufacture of binders, adhesives, surfactants, animal feed additives, and vanilla.

See Lignin.

LIME

*See **Calcium Oxide.***

LINEAR ALPHA OLEFINS

Linear hydrocarbons with a double bond at the end of the carbon chain are produced by the polymerization, or more correctly, the oligomerization of ethylene.

$$(n+2)CH_2=CH_2 \rightarrow CH_3CH_2(CH_2CH_2)_nCH=CH_2$$

Compounds with 6 to18 carbons are the most common alpha olefins (α-olefins) and Ziegler catalysts are used in this process. Certain olefins such as nonene (C_9) and dodecene (C_{12}) can also be made by cracking and dehydrogenation of n-paraffins, as practiced in the petrochemical section of a refinery.

Linear alpha olefins can be copolymerized with polyethylene to form linear low-density polyethylene (LLDPE) and 1-hexene (C_6) and 1-octene (C_8) are especially useful for this purpose. In addition, linear alpha olefins are used to make detergent alcohols, oxo alcohols for plasticizers, lubricants, lube oil additives, and surfactants are other important products from linear alpha olefins.

LIQUEFIED PETROLEUM GAS

Liquefied petroleum gas (LPG) is used for domestic and industrial heating, flame weeding, tobacco curing, grain drying, and in motor vehicles, as well as for the petrochemical industry.

The constituents of liquefied petroleum gas [propane ($CH_3CH_2CH_3$) and/or butane ($CH_3CH_2CH_2CH_3$)] occur as constituents of *wet* natural gas or crude oil or as a by-product from refining. For example, a natural gasoline plant treats raw *wet* natural gas through absorption by *washing* with gas oil and fractionating out the usable fraction.

Butane can be used for the manufacture of maleic acid (thence to maleic anhydride), from which tetrahydrofuran is made by hydrogenation. Liquefied petroleum gas is also a feedstock for aromatics production (Fig. 1).

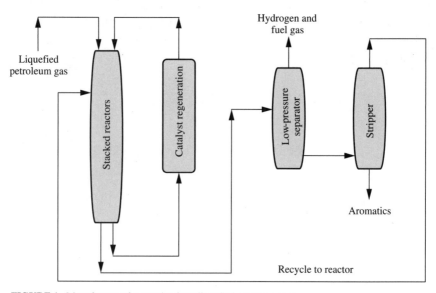

FIGURE 1 Manufacture of aromatics from liquefied petroleum gas.

LITHIUM SALTS

Spodumene is by far the most important lithium-containing ore and is used in the manufacture of lithium carbonate. Spodumene ore (beneficiated to 3 to 5% Li_2O) is converted from the alpha form to the beta form by heating to over 1000°C, since the alpha form is not attacked by hot sulfuric acid. The water-soluble lithium sulfate is leached out and reacted with sodium carbonate to yield lithium carbonate from which various salts are derived.

Lithium carbonate (Li_2CO_3), the most widely used of the compounds, is employed in the production of lithium metal and frits and enamels. Together with lithium fluoride (LiF), it serves as an additive for cryolite in the electrolytic pot line production of primary aluminum.

Lithium-base greases, especially the stearate, are efficient over an extremely wide temperature range up to 160°C. Lithium hydroxide (LiOH) is a component of the electrolyte in alkaline storage batteries and is employed in the removal of carbon dioxide in submarines and space capsules. Lithium bromide (LiBr) brine is used for air conditioning and dehumidification. Lithium hypochlorite (LiOCl) is a dry bleach used in commercial and home laundries. Lithium chloride (LiCl) is in demand for low-temperature batteries and for aluminum brazing. Other uses of lithium compounds include catalysts, glass manufacture, and, of course, nuclear energy.

LITHOPONE

Lithopone is a mixed zinc sulfide–barium sulfate brilliant white pigment that contains about 30% zinc sulfide. The original light sensitiveness of this pigment has been mitigated by purification and by the addition of such agents as polythionates and cobalt sulfate.

Lithopone is manufactured by a process (Fig. 1) in which barium sulfide solution is prepared by reducing barite ore ($BaSO_4$) with carbon and leaching the resulting mass.

$$BaSO_4 + 4C \rightarrow BaS + 4CO$$

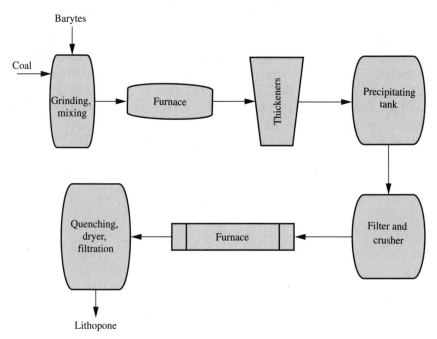

FIGURE 2 Manufacture of lithopone.

Scrap zinc or concentrated zinc ores are dissolved in sulfuric acid, the solution is purified, and the two solutions are reacted. A heavy mixed precipitate results that is 28 to 30% zinc sulfide and 72 to 70% barium sulfate.

$$ZnSO_4 + BaS \rightarrow ZnS + BaSO_4$$

This precipitate is not suitable for a pigment until it is filtered, dried, crushed, heated to a high temperature, and quenched in cold water. The second heating in a muffle furnace at 725°C produces crystals of the right optical size.

Lithopone is used in water-based paints because of its excellent alkali resistance. It is also used as a whitener and reinforcing agent for rubber and as a filler and whitener for paper.

MAGNESIUM

Magnesium occurs in seawater and in ores such as dolomite ($CaCO_3 \cdot MgCO_3$), magnesite ($MgCO_3$), and carnallite ($MgCl_2 \cdot KCl \cdot 6H_2O$).

Magnesium can be made by several methods (Fig. 1), but the most common method of manufacture is by the electrolytic process, as for example the electrolysis of magnesium chloride.

The magnesium chloride is obtained from saline solution, from brine, and from the reaction of magnesium hydroxide (from seawater or dolomite) with hydrochloric acid (Fig. 2). Electrolyzing magnesium chloride from seawater, using oyster shells for the lime needed is also an option. The oyster shells, which are almost pure calcium carbonate, are burned to lime, slaked, and mixed with the seawater, thus precipitating magnesium hydroxide. This magnesium hydroxide is filtered off and treated with hydrochloric acid prepared from the chlorine evolved by the cells to form magnesium chloride solution that is evaporated to solid magnesium chloride in direct-fired evaporators, followed by shelf drying. The chloride tends to decompose on drying and, after dehydrating, the magnesium chloride is fed to the electrolytic cells, where it is decomposed into the metal and chlorine gas. The internal parts of the cell act as the cathode, and 22 graphite anodes are suspended vertically from the top of the cell. The arrangement is very similar to the Downs sodium cell (Fig. 3).

Process	Source of chemical
Electrolysis of magnesium chloride	Magnesium oxide potash mining brine wells seawater
Ferrosilicon reduction process	Electric furnace blast furnace dolomite quarry

FIGURE 1   Process for the production of magnesium.

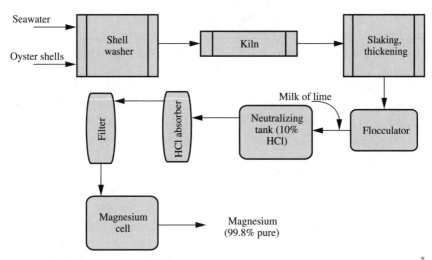

FIGURE 2   Production of magnesium from seawater.

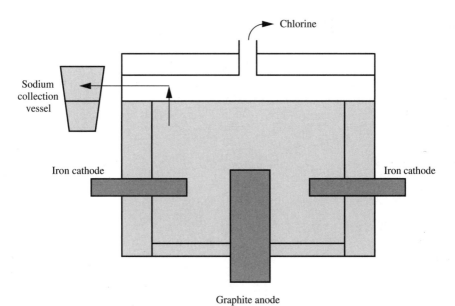

FIGURE 3   General schematic of the Downs cell.

Sodium chloride is added to the bath to lower the melting point and also increase the conductivity. The salts are kept molten by the electric current used to extract the magnesium plus external heat supplied by external gas-fired furnaces. The usual operating temperature is 710°C, which is sufficient to melt the magnesium (melting point 651°C). The molten magnesium is liberated at the cathode and rises to the bath surface, where troughs lead to the metal wells in front of the cell. The 99.9% pure magnesium metal is dipped out several times during the day, each dipperful containing enough metal to fill a 20-kg self-pelleting mold.

The silicothermic, or ferrosilicon, process involves mixing ground burned dolomite with ground ferrosilicon and fluorspar (eutectic) and pelletizing after which the pellets are charged into the furnace. High vacuum and heat (1170°C) are applied and the calcium oxide present in the burnt dolomite forms infusible calcium silicate that is removed from the retort.

$$12(MgO{\cdot}CaO) + 6FeSi_6 \rightarrow 12Mg + 6(CaO)_2SiO_2 + 6Fe$$

See **Magnesium Compounds**.

MAGNESIUM CARBONATE

Magnesium carbonate ($MgCO_3$) varies from dense material used in magnesite bricks to the very low density hydrated mixed carbonate-hydroxides [$4MgCO_3 \cdot Mg(OH)_2 \cdot 5H_2O$] and [$3MgCO_3 \cdot Mg(OH)_2 \cdot 3H_2O$] once employed for insulation. There are also other basic carbonates on the market with variations in adsorptive index and apparent density. Many of these are employed as fillers in inks, paints, and varnishes.

MAGNESIUM CHLORIDE

Magnesium chloride is made from hydrochloric acid and magnesium hydroxide.

$$Mg(OH)_2 + 2HCl \rightarrow MgCl_2 + 2H_2O$$

Magnesium chloride much resembles calcium chloride and has many of the same uses. In addition, it is used for production of magnesium, in ceramics, in the sizing of paper, and in the manufacture of oxychloride cement.

MAGNESIUM COMPOUNDS

The production of magnesium compounds by separation from aqueous solutions may be divided into four processes:

1. Manufacture from seawater without evaporation, using seawater and lime as the principal raw materials (Fig. 1).

2. Manufacture from bitterns or mother liquors from the solar evaporation of seawater for salt.

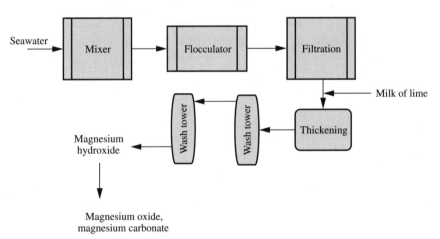

FIGURE 1 Production of magnesium compounds from seawater.

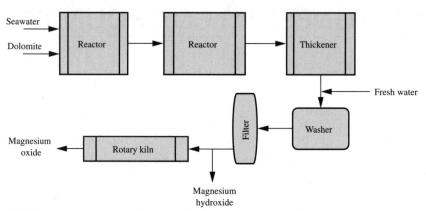

FIGURE 2   Production of magnesium hydroxide from seawater and dolomite.

3. Manufacture from dolomite ($MgCO_3$) and seawater (Fig. 2).

4. Manufacture from deep-well brines.

The production of magnesium compounds from seawater is made possible by the almost complete insolubility of magnesium hydroxide in water.

*See **Magnesium**.*

MAGNESIUM HYDROXIDE

*See **Magnesium Oxide**.*

MAGNESIUM OXIDE

Heating magnesium carbonate ($MgCO_3$) or magnesium hydroxide [$Mg(OH)_2$] produces magnesium oxide (MgO).

$$MgCO_3 \rightarrow MgO$$

$$Mg(OH)_2 \rightarrow MgO$$

Magnesium oxide has many uses, e.g., in the vulcanization of rubber, as a material for making other magnesium compounds, as an insulating material, as a refractory material, and as an abrasive.

See *Magnesium Compounds.*

MAGNESIUM PEROXIDE

Magnesium peroxide is manufactured by the reaction of magnesium sulfate and barium peroxide.

$$MgSO_4 + BaO_2 \rightarrow MgO_2 + BaSO_4$$

It is employed as an antiseptic and a bleaching agent.

MAGNESIUM SILICATE

Magnesium silicate ($MgSiO_3$ or $MgO \cdot SiO_2$) exists in two predominant forms–asbestos and talc. Asbestos is a magnesium silicate mixed with varying quantities of silicates of calcium and iron.

It is a fibrous noncombustible mineral and is used in the manufacture of many fireproof and insulating materials. Because of the cancer-causing characteristics of its fibers, government regulations have sharply reduced its use.

Talc is a pure magnesium silicate in the form of $3MgO \cdot 4SiO_2 \cdot H_2O$, found naturally in soapstone. It is employed as a filler in paper and plastics and in many cosmetic and toilet preparations.

MAGNESIUM SULFATE

Magnesium sulfate is prepared by the action of sulfuric acid on magnesium carbonate or hydroxide.

$$MgCO_3 + H_2SO_4 \rightarrow MgSO_4 + H_2O + CO_2$$

$$Mg(OH)_2 + H_2SO_4 \rightarrow MgSO_4 + 2H_2O$$

It is sold in many forms, one of which is the heptahydrate ($MgSO_4 \cdot 7H_2O$), long known as *Epsom salts*. The less pure material is used extensively as sizing and as a fireproofing agent.

MALATHION

Malathion, also known by the generic name *carbophos* (*o,o*-dimethyl phosphorodithioate or diethyl mercaptosuccinate), is one of the very popular, short-lived broad-spectrum insecticides for application to nearly all fruits, vegetables, field crops, dairy livestock, and household insects.

See Insecticides.

MALEIC ACID

*See **Liquefied Petroleum Gas**.*

MALEIC ANHYDRIDE

Maleic anhydride (melting point: 52.8°C, boiling point: 202°C, flash point: 110°C), formerly made from benzene, is now made from butane (Figs. 1 and 2) a switch in manufacturing method that was very rapid and complete.

Maleic anhydride is used in the manufacture of unsaturated polyester resins, copolymers, food additives, agricultural chemicals, and lube oil additives. Agricultural chemicals made from maleic anhydride include daminozide (Alar®), a growth regulator for apples that in 1989 was found to be carcinogenic because of a breakdown product, unsymmetrical dimethyl-hydrazine (UDMH).

$$HOOCCH_2CH_2CONHN(CH_3)CH_3 \qquad\qquad H_2NN(CH_3)CH_3$$
daminozide   UDMH

See Liquefied Petroleum Gas.

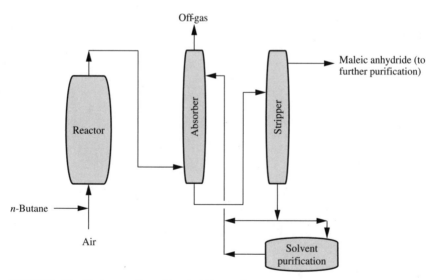

FIGURE 1   Fixed-bed process for maleic anhydride manufacture.

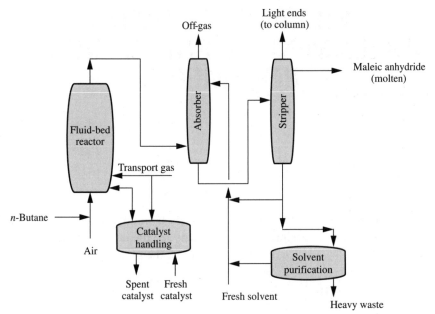

FIGURE 2 Fluid-bed process for maleic anhydride manufacture.

MELAMINE RESINS (MELAMINE-FORMALDEHYDE POLYMERS)

Melamine, having three amino groups and six labile hydrogen atoms, will also form thermoset resins with formaldehyde (melamine-formaldehyde polymers). The chemistry is similar to that for the urea resins.

*See **Urea Resins.***

MERCURY FULMINATE

See Explosives.

METALDEHYDE

Metaldehyde, a cyclic tetramer of acetaldehyde, is formed at temperatures below 0°C in the presence of dry hydrogen chloride or pyridine-hydrogen bromide. The metaldehyde crystallizes from solution and is separated from the paraldehyde by filtration.

Metaldehyde melts in a sealed tube at 246.2°C and sublimes at 115°C with partial depolymerization.

METHANE

Methane (CH_4, marsh gas, fire damp, melting point: $-182.6°C$, boiling point: $-161.4°C$, density: 0.415 at $-164°C$) is a colorless, odorless that is only very slightly soluble in water and moderately soluble in alcohol or ether. When ignited, the gas burns when ignited in air with a pale, faintly luminous flame. It forms an explosive mixture with air between gas concentrations of 5 and 13%.

Methane occurs as the principal constituent of natural gas and is often produced by separation from the other constituents of natural gas.

Alternatively, methane can be manufactured by the reaction of carbon monoxide and hydrogen in the presence of a nickel catalyst. Methane also is formed by reaction of magnesium methyl iodide (Grignard's reagent) in anhydrous ether with substances containing the hydroxyl group. Methyl iodide (bromide, chloride) is preferably made by reaction of methyl alcohol and phosphorus iodide (bromide, chloride).

Methane, as the major constituent of natural gas, is an extremely important raw material for numerous synthetic products. For most processes, it is not required to isolate and purify the methane, but the natural gas as received may be used.

Starting material	Reactant conditions	Product conditions	Reactant conditions	Product	Reactant	Product
Methane	Water heat	Synthesis gas	Water heat	Carbon dioxide, hydrogen	Nitrogen	Ammonia
		Synthesis gas	Catalyst heat	Methyl alcohol	Heat	Formaldehyde
				Methyl alcohol	i-butylene	MTBE
				Methyl alcohol	Terephthalic acid	Dimethyl terephthalate
				Methyl alcohol	Carbon monoxide	Acetic acid

FIGURE 1   Manufacture of chemicals from methane.

In addition to the preparation of synthesis gas, which is used so widely in various organic syntheses (Fig. 1), methane is reacted with ammonia in the presence of a platinum catalyst at a temperature of about 1250°C to form hydrogen cyanide:

$$CH_4 + NH_3 \rightarrow HCN + 3H_2$$

Methane also is used in the production of olefins on a large scale. In a controlled-oxidation process, methane is used as a raw material in the production of acetylene.

See Synthesis Gas.

METHYL ACETATE

See Acetic Acid.

METHYL ALCOHOL

Methyl alcohol (methanol, wood alcohol, CH_3OH; boiling point: 64.7°C, density: 0.7866, flash point: 110°C) is a colorless, mobile liquid with a mild characteristic odor (and narcotic properties) that is miscible in all proportions with water, ethyl alcohol, or ether. When ignited, methyl alcohol burns in air with a pale blue, transparent flame, producing water and carbon dioxide. The vapor forms an explosive mixture with air. The upper explosive limit is 36.5% and the lower limit is 6.0% by volume in air.

Before 1926 all methyl alcohol was made by distillation of wood. Now methanol is produced by synthetic methods.

Methyl alcohol is obtained from synthesis gas under appropriate conditions (Fig. 1) or by the oxidation of methane (Fig. 2). This includes zinc, chromium, manganese, or aluminum oxides as catalysts, 300°C, 250 to

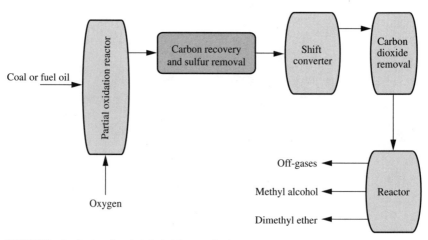

FIGURE 1　Production of methyl alcohol from synthesis gas.

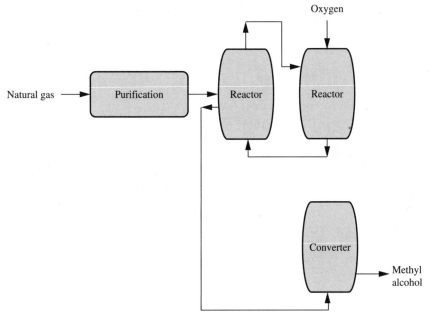

FIGURE 2 Production of methyl alcohol from methane (natural gas).

300 atm (3000 to 5000 psi), and most importantly a 1:2 ratio of carbon monoxide (CO) to hydrogen (H_2). Copper oxide catalysts require lower temperatures and pressures, usually 200 to 300°C and 50 to 100 atm (750 to 1500 psi). A 60 percent yield of methanol is realized.

$$3CH_4 + 2H_2O + CO_2 \rightarrow 4CO + 8H_2$$

$$CO + 2H_2 \rightarrow CH_3OH$$

The methanol can be condensed and purified by distillation. Unreacted synthesis gas is recycled. Other products include higher boiling alcohols and dimethyl ether (CH_3OCH_3).

Methyl alcohol is also obtained by the oxidation of methane using natural gas as the feedstock (Fig. 2).

See Synthesis Gas.

METHYLAMINES

Methylamine (CH_3NH_2, boiling point: –6.3°C), dimethylamine [$(CH_3)_2NH$, boiling point: 7°C], and trimethylamine [$(CH_3)_3N$, boiling point: 3°C] are manufactured by the reaction of methanol with ammonia (Fig. 1).

$$CH_3OH + NH_3 \rightarrow CH_3NH_2 + H_2O$$

$$2CH_3OH + NH_3 \rightarrow (CH_3)_2NH + 2H_2O$$

$$3CH_3OH + NH_3 \rightarrow (CH_3)_3N + 3H_2O$$

The ratios of the reactants can be adjusted to produce the desired product mix.

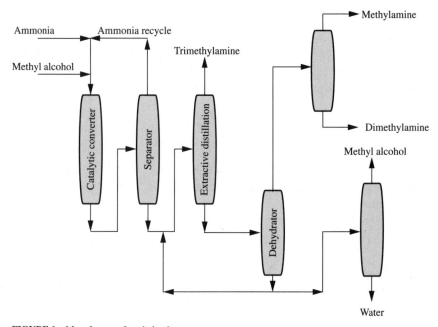

FIGURE 1 Manufacture of methylamines.

METHYL CHLORIDE

The major method for the production of methyl chloride (melting point: −97.1°C, boiling point: −24.2°C, density: 0.9159) is by the reaction of methanol and hydrogen chloride, with the aid of a catalyst, in either the vapor or liquid phase.

$$CH_3OH + HCl \rightarrow CH_3Cl + H_2O$$

Methyl chloride is also manufactured by the chlorination of methane.

$$CH_4 + Cl_2 \rightarrow CH_3Cl + HCl$$

Methyl chloride is used for the production of silicones, agricultural chemicals, methylcellulose, and quaternary amines.

METHYLENE CHLORIDE

Methylene chloride (methylene dichloride, dichloromethane, melting point: $-95.1°C$, boiling point: $40°C$, density: 1.3266) is produced by the chlorination of methyl chloride, which in turn is made by the chlorination of methane.

$$CH_4 + Cl_2 \rightarrow CH_3Cl + HCl$$

$$CH_3Cl + Cl_2 \rightarrow CH_2Cl_2 + HCl$$

The main uses of methylene chloride are in paint remover, aerosols, chemical processing, urethane foam blowing agents, metal degreasing, and electronics.

METHYLENE DIPHENYL DIISOCYANATE

Methylene diphenyl diisocyanate (MDI) is produced by the condensation of aniline with formaldehyde followed by reaction with phosgene.

$$C_6H_5-NH_2 + CH_2=O \rightarrow CH_2(C_6H_4-NH_2)_2$$

$$CH_2(C_6H_4-NH_2)_2 + COCl_2 \rightarrow CH_2(C_6H_4-N=C=O)_2$$

Methylene diphenyl diisocyanate is used to manufacture polyurethane foams for construction, refrigeration, and packaging.

METHYL ETHYL KETONE

Methyl ethyl ketone (MEK; boiling point: 769.6°C, density: 0.8062, flash point: –6°C) is an important coating solvent for many polymers and is made by the sulfation and hydration of 1 or 2-butene to *sec*-butyl alcohol, which is then dehydrogenated to the ketone (Fig. 1).

$$CH_3CH_2CH=CH_2 \rightarrow CH_3CH_2CH(OH)CH_3$$

$$CH_3CH=CHCH_3 \rightarrow CH_3CH_2CH(OH)CH_3$$

$$CH_3CH_2CH(OH)CH_3 \rightarrow CH_3CH_2C(=O)CH_3$$

Butane oxidation is another route to methyl ethyl ketone.

$$CH_3CH_2CH_2CH_3 \rightarrow CH_3CH_2COCH_3$$

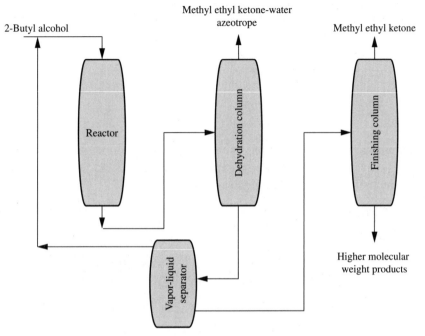

FIGURE 1   Manufacture of methyl ethyl ketone.

METHYL METHACRYLATE

Methyl methacrylate (melting point: –48°C, boiling point 100°C, density: 0.9394, flash point: 9°C) is produced by the acetone cyanohydrin process in which the acetone cyanohydrin (from the reaction of acetone with hydrogen cyanide, *q.v.*) is reacted with sulfuric acid to yield methacrylamide sulfate, which is further hydrolyzed and esterified. The process is continuous.

$$CH_3C(OH \cdot CN)CH_3 + H^+ \rightarrow CH_2=C(CH_3)C(=O)NH_3^+$$

$$CH_2=C(CH_3)C(=O)NH_3^+ + CH_3OH + HSO_4^- \rightarrow$$

$$CH_2=C(CH_3)C(=O)OCH_3 + NH_4SO_4$$

Methyl methacrylate is also manufactured by oxidation of *iso*-butene or *t*-butyl alcohol.

$$(CH_3)_2C=CH_2 \rightarrow CH_2=C(CH_3)C(=O)OCH_3$$

$$(CH_3)_3COH \rightarrow CH_2=C(CH_3)C(=O)OCH_3$$

Other routes to methyl methacrylate include starting with *t*-butyl alcohol and ethylene:

$$(CH_3)_3COH + [O] \rightarrow CH_2=C(CH_3)CH=O$$

$$CH_2=C(CH_3)CH=O + [O] \rightarrow CH_2=C(CH_3)CO_2H$$

$$CH_2=C(CH_3)CO_2H + CH_3OH \rightarrow CH_2=C(CH_3)C(=O)OCH_3$$

or

$$CH_2=CH_2 + H_2 + CO \rightarrow CH_3CH_2CH=O$$

$$CH_3CH_2CH=O + HCH=O \rightarrow CH_2=C(CH_3)CH=O$$

$$CH_2=C(CH_3)CH=O + \{O] \rightarrow CH_2=C(CH_3)CO_2H$$

$$CH_2=C(CH_3)CO_2H + CH_3OH \rightarrow CH_2=C(CH_3)C(=O)OCH_3$$

Methyl methacrylate is polymerized to poly(methyl methacrylate), which is used in cast and extruded sheet, molding powder and resins, surface coatings, impact modifiers, and emulsion polymers.

METHYL TERTIARY BUTYL ETHER

Methyl tertiary butyl ether (methyl-*t*-butyl ether, MTBE; boiling point: 55°C, flash point: –30°C) has excited considerable interest because it is a good octane enhancer for gasoline (it blends as if it had a research octane number of 115 to 135). It also offers a method of selectively removing *iso*-butylene from a mixed C_4 stream, thus enabling the recovery of high-purity butene-1. Furthermore, methyl tertiary butyl ether can be isolated, then cracked to yield highly pure *iso*-butylene and methanol.

The reaction for making methyl-*t*-butyl ether proceeds quickly and highly selectively by reacting a mixed butene-butane fraction with methyl alcohol in the liquid phase on a fixed bed of an acidic ion-exchange resin catalyst (Fig. 1).

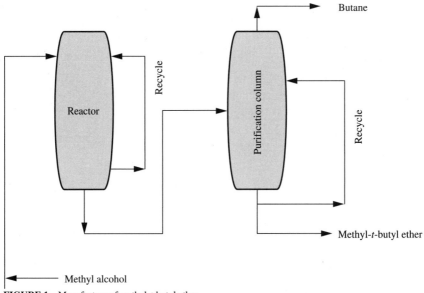

FIGURE 1 Manufacture of methyl-*t*-butyl ether.

$$CH_3OH + (CH_3)_2C{=}CH_2 \rightarrow CH_3OC(CH_3)_3$$

Reactor effluent is distilled, giving methyl-*t*-butyl ether of 99% purity. A few percent of *iso*-butylene remains unchanged and can be scavenged by use of a second unit.

METHYL VINYL ETHER

*See **Vinyl Ethers.***

MOLYBDENUM COMPOUNDS

Molybdenum occurs naturally as the mineral *molybdenite,* which, by roasting, produces molybdenum trioxide (up to 90% purity).

Molybdenum disulfide is dispersed in greases and oils for lubrication; in volatile carriers it is used to form dry coatings of lubricant. Sodium molybdate is an especially effective corrosion inhibitor on aluminum surfaces and is dissolved in cooling solutions to protect aluminum motor blocks in automobiles.

Molybdenum salts used as catalysts include cobalt molybdate for hydrogen treatment of petroleum stocks for desulfurization, and phosphomolybdates to promote oxidation. Compounds used for dyes are sodium, potassium, and ammonium molybdates. With basic dyes, phosphomolybdic acid is employed. The pigment known as *molybdenum orange* is a mixed crystal of lead chromate and lead molybdate. Sodium molybdate, or molybdic oxide, is added to fertilizers as a beneficial trace element. Zinc and calcium molybdate serve as inhibitory pigments in protective coatings arid paint for metals subjected to a corrosive atmosphere. Compounds used to produce better adherence of enamels are molybdenum trioxide and ammonium, sodium, calcium, barium, and lead molybdates.

MONOSODIUM GLUTAMATE

Monosodium glutamate (MSG) is an important flavoring agent, yet has no flavor of its own but it does accentuate the flavors of food in which it is used. Glutamic acid exists in three forms, but only the monosodium salt of L-glutamic acid has a flavor-accentuating capacity. Glutamic acid is a constituent of all common proteins and is produced by a process in which the principal steps are (Fig. 1): (1) concentration and collection of the filtrate, (2) hydrolysis, usually with caustic soda, (3) neutralization and acidification of the hydrolysate, (4) partial removal of the inorganic salts, and (5) crystallization, separation, and purification of the glutamic acid.

L-Glutamic acid can be obtained directly from fermentation of carbohydrates with *Micrococcus glutarnicus* or *Brevihacterium divaricatum.*

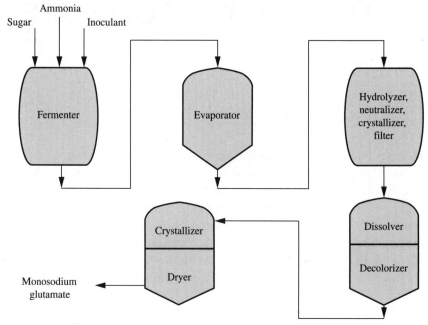

FIGURE 1   Manufacture of monosodium glutamate.

MORPHINE

Morphine ($C_{17}H_{19}NO_3 \cdot x H_2O$, melting point: 253°C) is a white powder that is derived from opium which is the dried juice obtained from unripe capsules of the poppy plant (*Papaver somniferum*), cultivated in various regions of the world.

The opium poppy is an annual. When the petals drop from the white flowers, the capsules are cut. The juice exudes and hardens, forming a brownish mass that is crude opium. It contains a total of about 20 narcotics, including morphine.

Morphine (about 11%) and codeine (about 1%) are extracted, along with many of the other alkaloids occurring in opium, by mixing sliced opium balls or crushed dried opium with lime water and removing the alkaloids by countercurrent aqueous techniques. Other solvents are also used, for instance, acetone and acetic acid or acidulated water. The crude morphine alkaloid is precipitated with ammonium chloride, purified by crystallization of one of its inorganic salts from water, and centrifuged; crystallization is repeated if necessary. The purified sulfate or hydrochloride is converted into the alkaloid by ammonia precipitation. If a still further purified alkaloid is needed, it can be prepared by crystallization from alcohol. Otherwise the morphine alkaloid is dissolved in water with sulfuric acid and crystallized in large cakes from which the mother liquor is drained and then sucked off. The sulfate is dried and cut into convenient sizes for the manufacturing pharmacist to compound it, make it into tablets, or otherwise facilitate its use by physicians.

Morphine has been, and remains, an important drug even though codeine is used to a larger extent than morphine and, while its analgesic action is only one-sixth of morphine, it is employed for its antitussive effect, as a cough repressant.

The discovery of morphine's analgesic activity in 1806 started a long series of studies of the alkaloids from the opium poppy, including morphine's first correctly postulated structure in 1925 and its total synthesis in 1952. The depressant action of the morphine group is the most useful property, resulting in an increased tolerance to pain, a sleepy feeling, a lessened perception to external stimuli, and a feeling of well being. Respiratory depression and addiction are its serious drawbacks.

The important structure-activity relationships that have been defined are: (1) a tertiary nitrogen, the group on the nitrogen being small; (2) a central carbon atom, of which none of the valences is connected to hydrogen; (3) a phenyl group connected to the central carbon; and (4) a two-carbon chain separating the central carbon from the nitrogen.

The *codeine* that occurs naturally in small amounts in opium is isolated from the aqueous morphine alkaloid mother liquors by immiscible extraction with a nonaqueous solvent. Dilute sulfuric acid is employed to extract the codeine sulfate from the nonaqueous solvent. This solution is evaporated, crystallized, and recrystallized. The alkaloid is precipitated from a sulfate solution by alkali and purified, if necessary, by alcoholic crystallization. It is converted into the phosphate by solution in phosphoric acid, evaporation, crystallization, centrifugation, and drying.

Codeine is the methyl ether of morphine and therefore is also prepared from morphine by methylating the phenolic hydroxyl group with diazomethane, dimethyl sulfate, or methyl iodide.

Since the source of the natural alkaloids is opium, all narcotics whose actions resemble those of morphine are sometimes referred to as *opiates*. Semisynthetic agents are usually made by altering the morphine molecule, and include such agents as ethylmorphine (Dionin), dihydromorphinone (Dilaudid), and methyldihydromorphinone (Metopon). Synthetic narcotics include agents with a wide variety of chemical structures. Some of the important synthetic agents are meperidine (piperidine-type), levorphanol (morphinian-type), methadone (aliphatic-type), phenaxocine (benzmorphan-type), and their derivatives.

Heroin is diacetylmorphine (diamorphine hydrochloride) and is prepared by the action of acetic anhydride on morphine. It possesses four times the analgesic effect of morphine, but has considerably less depressant effect. Addiction is common, the drug being taken in the form of snuff, or by injection.

NAPHTHA

Naphtha is a generic term applied to a refined or partly refined petroleum product (Fig. 1) that distills below 240°C (465°F) under standardized distillation conditions.

Naphtha is valuable for solvents because of its good dissolving power. The wide range of naphthas available, from the ordinary paraffinic straight-run to the highly aromatic types, and the varying degree of volatility possible offer products suitable for many uses.

Naphtha is divided into two main types, aliphatic and aromatic. Aliphatic naphtha is composed of paraffinic hydrocarbons and cycloparaffins (naphthenes), and may be obtained directly from crude petroleum by distillation. Aromatic naphtha contains aromatics, usually alkyl-substituted benzene, and is very rarely, if at all, obtained from petroleum as straight-run materials; often reforming is necessary (Fig. 2).

In general, naphtha may be manufactured by any one of several methods, including:

1. Fractionation of straight-run, cracked, and reforming distillates, or even fractionation of crude petroleum
2. Solvent extraction
3. Hydrogenation of cracked distillates
4. Polymerization of unsaturated compounds (olefins)
5. Alkylation processes

In fact, the naphtha may be a combination of product streams from more than one of these processes.

The most common method of naphtha preparation is distillation. Depending on the design of the distillation unit, either one or two naphtha steams may be produced: (1) a single naphtha with an end point of about 205°C (400°F) and similar to straight-run gasoline, or (2) this same fraction divided into a light naphtha and a heavy naphtha. The end point of the light naphtha is varied to suit the subsequent subdivision of the naphthas into narrower boiling fractions and may be of the order of 120° C (250° F).

Process	Primary product	Secondary process	Secondary product
Atmospheric distillation	Naphtha		Light naphtha
			Heavy naphtha
	gas oil	Catalytic cracking	Naphtha
	gas oil	Hydrocracking	Naphtha
Vacuum Distillation	gas oil	Catalytic cracking	Naphtha
		Hydrocracking	Naphtha
	residiuum	Coking	Naphtha
		Hydrocracking	Naphtha

FIGURE 1   Naphtha production in a petroleum refinery.

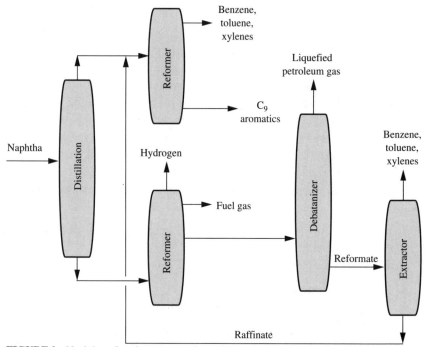

FIGURE 2   Naphtha reforming to aromatic products.

Before the naphtha is redistilled into a number of fractions with boiling ranges suitable for aliphatic solvents, the naphthas are usually treated to remove sulfur compounds, as well as aromatic hydrocarbons, which are present in sufficient quantity to cause an odor. Aliphatic solvents that are specially treated to remove aromatic hydrocarbons are known as *deodorized solvents.* *Odorless solvent* is the name given to heavy alkylate used as an aliphatic solvent, which is a by-product in the manufacture of aviation alkylate.

Naphtha that is either naturally sweet (no odor) or has been treated until sweet is subdivided into several fractions in efficient fractional distillation towers, frequency called *column steam stills*. A typical arrangement consists of primary and secondary fractional distillation towers and a stripper. Heavy naphtha, for example, is heated by a steam heater and passed into the primary tower, which is usually operated under vacuum. The vacuum permits vaporization of the naphtha at the temperatures obtainable from the steam heater.

The primary tower separates the naphtha into three parts:

1. The unwanted heavy ends, which are removed as a bottom product and sent to cracking coil stock
2. A side stream product of narrow boiling range, which after passing through the stripper may be suitable for the aliphatic solvent Varsol®
3. An overhead product, which is pumped to the secondary tower

The overhead product from the primary tower is divided into an overhead and a bottom product in the secondary tower, which operates under a partial vacuum with steam injected into the bottom of the tower to assist in the fractionation. The overhead and bottom products are finished aliphatic solvents, or if the feed to the primary tower is light naphtha instead of heavy naphtha, other aliphatic solvents of different boiling ranges are produced.

Several methods, involving solvent extraction or destructive hydrogenation, can accomplish the removal of aromatic hydrocarbons from naphtha. By destructive hydrodegation methods, aromatic hydrocarbon rings are first ruptured and then saturated with hydrogen, which converts aromatic hydrocarbons into the odorless, straight-chain paraffinic hydrocarbons required in aliphatic solvents.

Extractive distillation, that is, fractional distillation in the presence of a solvent, is used to recover aromatic hydrocarbons from, say, reformate fractions in the following manner. By means of preliminary distillation in a 65-tray prefractionator, a fraction containing a single aromatic can be separated from the reformate, and this aromatic concentrate is then pumped to an extraction distillation tower near the top, and aromatic concentrate enters near the bottom. A reboiler in the extractive distillation tower induces the aromatic concentrate to ascend the tower, where it contacts the descending solvent.

The solvent removes the aromatic constituents and accumulates at the bottom of the tower; the nonaromatic portion of the concentrate leaves the top of the tower and may contain about 1 percent of the aromatics. The solvent and dissolved aromatics are conveyed from the bottom of the extractive distillation tower to a solvent stripper, where fractional distillation separates the aromatics from the solvent as an overhead product. The solvent is recirculated to the extractive distillation tower, whereas the aromatic stream is treated with sulfuric acid and clay to yield a finished product of high purity.

Silica gel is an adsorbent for aromatics and has found use in extracting aromatics from refinery streams. Silica gel is manufactured amorphous silica that is extremely porous and has the property of selectively removing and holding certain chemical compounds from mixtures. For example, silica gel selectively removes aromatics from a petroleum fraction, and after the nonaromatic portion of the fraction is drained from the silica gel, the adsorbed aromatics are washed from the silica gel by a stripper (or desorbent). Depending on the kind of feedstock, xylene, kerosene, or pentane may be used as the desorbent.

The main uses of petroleum naphtha fall into the general areas of

1. Solvents (diluents) for paints, for example
2. Dry-cleaning solvents
3. Solvents for cutback asphalt
4. Solvents in the rubber industry
5. Solvents for industrial extraction processes

Turpentine, the older, more conventional solvent for paints, has now been almost completely replaced with the discovery that the cheaper and more abundant petroleum naphthas are equally satisfactory. The differences in application are slight; naphthas cause a slightly greater decrease in viscosity when added to some paints than does turpentine, and depending on the boiling range, they may show some differences in evaporation rates.

Naphthas are used in the rubber industry for dampening the play and tread stocks of automobile tires during manufacture to obtain better adhesion between the units of the tire. They are also consumed extensively in making rubber cements (adhesives) or are employed in the fabrication of rubberized cloth, hot-water bottles, bathing caps, gloves, overshoes, and toys. These cements are solutions of rubber and were formerly made with benzene, but petroleum naphtha is now preferred because of its less toxic character.

Naphthas are used for extraction on a fairly wide scale. They are applied in extracting residual oil from castor beans, soybeans, cottonseed, and wheat germ and in the recovery of *grease* from mixed garbage and refuse. The solvent employed in these cases is a hexane cut, boiling from about 65 to 120°C (150 to 250°F). When the oils recovered are of edible grade or intended for refined purposes, stable solvents completely free of residual odor and taste are necessary, and straight-run streams from low-sulfur, paraffinic crude oils are generally satisfactory.

The recovery of wood resin by naphtha extraction of the resinous portions of dead trees of the resin-bearing varieties or stumps, for example, is also used in the wood industry. The chipped wood is steamed to distill out the resinous products recoverable in this way and then extracted with a naphtha solvent, usually a well-refined, low-sulfur, paraffinic product boiling from, say, 95 to 150°C (200 to 300°F).

Naphthas are also employed as solvents in the manufacture of printing inks, leather coatings, diluents for dyes, and degreasing of wool fibers, polishes, and waxes, as well as rust- and waterproofing compositions, mildew-proofing compositions, insecticides, and wood preservatives.

NAPHTHALENE

Naphthalene (melting point: 80.3°C, density: 1.175, flash point: 79°C) is very slightly soluble in water but is appreciably soluble in many organic solvents such as 1,2,3,4-tetrahydronaphthalene (tetralin), phenols, ethers, carbon disulfide, chloroform, benzene, coal-tar naphtha, carbon tetrachloride, acetone, and decahydronaphthalene (decalin).

Naphthalene is produced from coal tar. In the *coal tar process*, coal tar is processed through a tar-distillation step where approximately the first 20 wt % of distillate (*chemical oil*) is removed. The chemical oil, which contains practically all the naphthalene present in the tar, is reserved for further processing, and the remainder of the tar is distilled further to remove additional creosote oil fractions until a coal-tar pitch of desirable consistency and properties is obtained. The chemical oil is processed to remove the tar acids by contacting with dilute sodium hydroxide and, in a few cases, is next treated to remove tar bases by washing with sulfuric acid.

Crude naphthalene product is obtained by fractional distillation of the tar acid–free chemical oil, and the distillation may be accomplished in either a batch process or a continuous process (Fig. 1). The tar acid–free chemical oil is charged to the system, where most of the low boiling components such as benzene, xylene, and toluene, are removed in the light-solvent column. The chemical oil next is fed to the solvent column, which is operated under vacuum, where a product containing the prenaphthalene components is taken overhead. This product, which is called *coal-tar naphtha* or *crude heavy solvent*, typically has a boiling range of approximately 130 to 200°C and is used as a general solvent and as a feedstock for hydrocarbon-resin manufacture because of its high content of compounds such as indene and coumarone.

The naphthalene-rich bottoms from the solvent column then are fed to the naphthalene column where a naphthalene product (95% naphthalene) is produced. The naphthalene column is operated at near atmospheric pressure to avoid difficulties that are inherent to vacuum distillation of this product, for example, naphthalene-filled vacuum jets and lines. A side stream that is rich in methylnaphthalenes may be taken near the bottom of the naphthalene column.

The main impurity in crude 78°C coal-tar naphthalene is sulfur that is present in the form of thionaphthene (1 to 3%). Methyl- and dimethyl-naphthalenes also are present (1 to 2%) with lesser amounts of indene, methylindenes, tar acids, and tar bases.

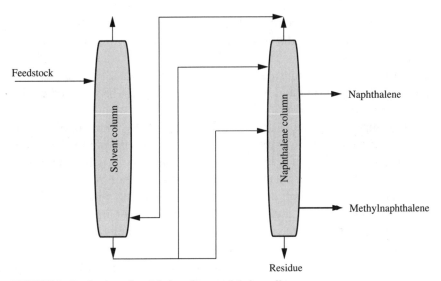

FIGURE 1   Production of naphthalene from naphthalene oil.

NATURAL GAS

Natural gas (predominantly methane, CH_4, with ethane, propane, and butane) is a fuel gas, and it is also an important chemical raw material for various syntheses.

In addition to the industrially valuable hydrocarbons, natural gas contains undesirable water and hydrogen sulfide that must be removed before shipping to the consumer.

Four important methods are employed for the dehydration of gas: compression, treatment with drying substances, adsorption, and refrigeration. A plant for water removal by compression consists of a gas compressor, followed by a cooling system to remove the water vapor by condensation. The treatment of gas with drying substances has found widespread usage in this country. Glycols are used most widely for this purpose because of their high affinity for water, chemical stability, low foaming, and low solvent action for natural gas. For water dew points in the range of −90 to − 100°C, molecular sieves are used in many plants. The beds are regenerated by countercurrent flow of hot gases (230 to 290°C).

Other drying agents are activated alumina and bauxite, silica gel, sulfuric acid, and concentrated solutions of calcium chloride or sodium thiocyanate. Plants of this type usually require a packed tower for countercurrent treatment of the gas with the reagent, together with a regenerator for the dehydrating agent.

Hydrogen sulfide and other sulfur compounds are objectionable in natural gas because they cause corrosion and also form air-polluting compounds during combustion. Carbon dioxide in the gas is objectionable because it lowers the heating value of the gas.

Monoethanolamine is the oldest and probably still the most widely used solvent (Fig. 1). For desulfurization of natural gas, a 10 to 30% aqueous solution of monoethanolamine is normally used and a variety of solvents are available that vary in solvent selectivity for absorption of hydrogen sulfide and carbon dioxide, and this property, as well as the composition of the impurities in the gas being treated, frequently determines the choice of solvent. Some of

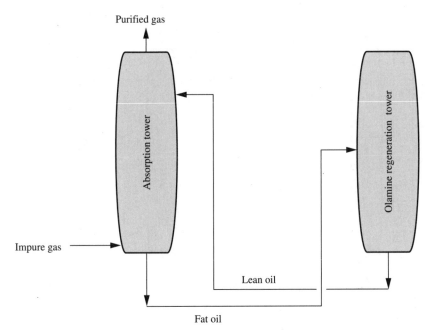

FIGURE 1 Gas purification using an olamine.

the solvents also have a high affinity for higher hydrocarbons, and this is a disadvantage if the gas contains an appreciable quantity of these valuable compounds. If simultaneous dehydration and desulfurization is desired, the gas may be scrubbed with a solution of amine, water, and glycol. Solution compositions for this purpose are from 10 to 36% monoethanolamine, 45 to 85% diethylene glycol, and the remainder water.

One of the newest commercial methods of sweetening gas is the use of membranes. This separation works on the principle that there are different rates of permeation through a membrane for different gases.

Membrane materials used are polysulfone, polystyrene, Teflon, and various rubbers. This type of separation possesses many advantages over other types of gas separation, e.g., mild operating conditions, lower energy consumption, low capital cost, and economic operation at both low and high flow rates.

Natural gas with a high nitrogen content can be upgraded by a cryogenic process that dries feed gas at 4.9 MPa and cools it to 185 K. The natural gas is vaporized, and both this and the separated nitrogen gas leave the system via heat exchangers against incoming gas.

Natural gas can also be separated into its hydrocarbon constituents (Fig. 2), thereby producing feedstocks for petrochemical processes.

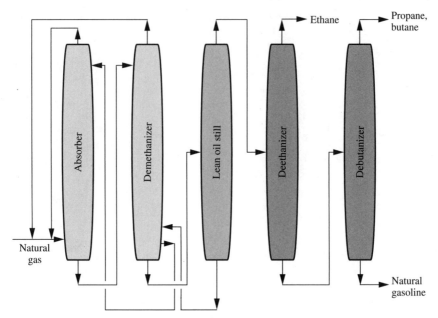

FIGURE 2   Separation of natural gas into hydrocarbon constituents.

NATURAL GAS (SUBSTITUTE)

Substitute natural gas is produced by a series of reactions in a variety of gasifiers that use coal as the feedstock (Fig. 1). The low- and medium-heat syngas (carbon monoxide and hydrogen) so produced are converted to a high-heat-content gas similar to natural gas:

$$C + H_2O \rightarrow CO + H_2$$

$$CO + H_2O \rightarrow CO_2 + H_2$$
$$C + CO_2 \rightarrow 2CO$$

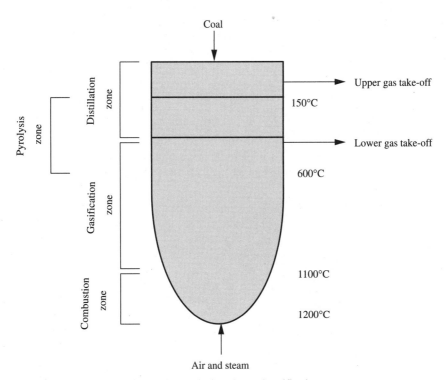

FIGURE 1 Production of natural gas substitute by coal gasification.

At sufficiently high pressures, the hydrogen produced by the first two reactions will hydrogenate some of the carbon to yield methane.

$$C + 2H_2 \rightarrow CH_4$$
$$CO + 3H_2 \rightarrow CH_4 + H_2O$$

The gas thus produced was known originally as *synthetic natural gas*, but it can be argued that the correct name is *substitute natural gas*.

The sulfur and carbon dioxide are removed from the gas before it is methanated. The operating temperature can vary from 800 to about 1650°C with a pressure that will vary depending upon the gasifier and the desired product. The higher pressure and lower temperature result in the formation of a larger amount of methane.

Naphtha and other oils can be converted to substitute natural gas. The processes use mix naphtha with steam in a 1:2 ratio and gasify the mixture. The gas produced is methanated by the reaction of the carbon oxides with the hydrogen present. Purification requires the removal of any residual carbon dioxide.

See Naphtha.

NEON

*See **Rare Gases.***

NICOTINE

Nicotine (melting point: −79°C, boiling point: 246.7°C, density: 1.0097) is a volatile alkaloid obtained by treating tobacco waste with aqueous alkali, followed by steam distillation. Most nicotine is converted to the less volatile sulfate and sold as a 40% solution.

In addition to being an active ingredient in tobacco smoke and having a pharmacological effect on humans, nicotine (in solution) is effective against aphids, thrips, and leaf hoppers and can he used as a fumigant.

NICOTINIC ACID AND NICOTINAMIDE

Nicotinic acid (melting point: 236°C, density: 1.473) and nicotinamide (melting point: 129°C, density: 1.400) are known as *niacin* and *niacinamide* in the food industry. Niacin is the most stable of all vitamins and is essential to humans and animals for growth and health. Niacin and niacinamide are nutritionally equivalent, and compete with one another.

For production of niacinamide in the past, methylethylpyridine was oxidized with nitric acid to yield niacin, and β-picoline was treated with air and ammonia to produce the nitrile that was then hydrolyzed to niacinamide. A more modern process can produce both niacin and niacinamide from a single feedstock, either β-picoline or 2-methyl-5-ethylpyridine by oxidative ammonolysis, a combination of oxidation and amination.

Either niacin or niacinamide can be selectively isolated from the hydrolysis by varying the hydrolysis time and nitrile concentration. A higher hydrolysis temperature favors production of niacin.

NITRIC ACID

Nitric acid (HNO_3; freezing point: –41.6°C, boiling point: 86°C, density: 1.503 at 25°C) is a colorless, highly corrosive liquid and a very powerful oxidizing agent that in the highly pure state is not entirely stable and must be prepared from its azeotrope by distillation with concentrated sulfuric acid. Nitric acid gradually yellows because of decomposition to nitrogen dioxide. Solutions containing more than 80% nitric acid are called *fuming nitric acids*.

Reagent-grade nitric acid is a water solution containing about 68% by weight nitric acid. This strength corresponds to the constant-boiling mixture of the acid with water, which is 68.4% by weight nitric acid and boils at 121.9°C. Nitric acid is completely miscible with water and forms a monohydrate ($HNO_3 \cdot H_2O$, melting point: $-38°C$) and a dihydrate ($HNO_3 \cdot 2H_2O$, melting point: $-18.5°C$).

For many years nitric acid was made by the reaction of sulfuric acid and saltpeter, but this method is no longer used.

$$NaNO_3 + H_2SO_4 \rightarrow NaHSO_4 + HNO_3$$

Nitric acid is now manufactured by combusting ammonia in air in the presence of a (platinum or other noble metal) catalyst, and the nitrogen oxides thus formed are oxidized further and absorbed in water to form nitric acid.

$$NH_3 + 2O_2 \rightarrow HNO_3 + H_2O$$

In this process (Fig. 1), the reactor contains a rhodium-platinum catalyst (2 to 10% rhodium) as wire gauzes in layers of 10 to 30 sheets at 750 to 920°C, 100 psi, and a contact time of 3×10^{-4} second. After cooling, the product gas enters the absorption tower with water and more air to oxidize the nitric oxide and hydrate it to nitric acid in water. Waste gases contain nitric oxide or nitrogen dioxide, and these are reduced with hydrogen or methane to ammonia or nitrogen gas. Traces of nitrogen oxides can be

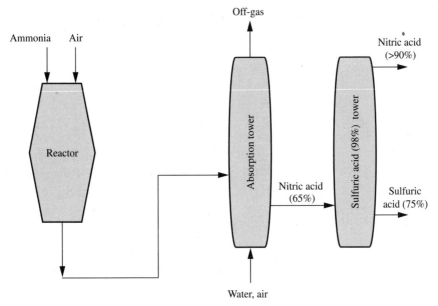

FIGURE 1 Manufacture of nitric acid.

expelled. Concentration of the nitric acid in a silicon-iron or stoneware tower containing 98% sulfuric acid will give 90% nitric acid off the top and 70 to 75% sulfuric acid as the bottoms. This last step is necessary because simple distillation of nitric acid is not applicable; it forms an azeotrope with water at 68% acid. An alternative drying agent is magnesium nitrate, which can concentrate the acid to 100% nitric acid.

Manufactured acid contains some nitrous acid (HNO_2) when the concentration is between 20 and 45% nitric acid and dissolved nitrogen tetraoxide (N_2O_4); when the concentration is over 55%. The oxidation potential of 20 to 45% acid stabilizes trivalent nitrogen (HNO_2), over 55% stabilizes tetravalent nitrogen (N_2O_4).

Nitric acid is predominantly used for fertilizer manufacture. It also finds use in the manufacture of adipic acid, nitroglycerin, nitrocellulose, ammonium picrate, trinitrotoluene, nitrobenzene, silver nitrate, and various isocyanates.

NITROBENZENE

Nitrobenzene (melting point: 5.9°C, boiling point: 210.9°C, density: 1.199, flash point: 88°C) is made by the direct nitration of benzene using a nitric acid–sulfuric acid mixture (Fig.1), usually in a cast-iron or steel kettle.

$$C_6H_6 + HNO_3 + H+ \rightarrow C_6H_5NO_2 + H_2O$$

The majority of the nitrobenzene is used for the production of aniline with a minor amount being used for the production of acetaminophen.

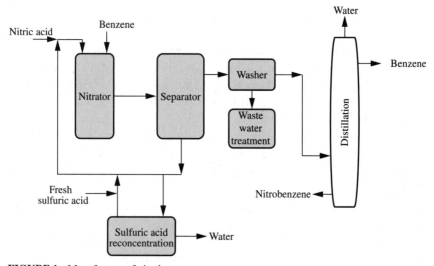

FIGURE 1　Manufacture of nitrobenzene.

NITROCELLULOSE

Nitrocellulose is a generic name for a product produced by the reaction of cellulosic materials with nitric acid and thereby containing the nitro group.

The commercial manufacture of cellulose nitrate involves treatment of cellulose with a mixture of nitric and sulfuric acids:

$$C_6H_7O_2(OH)_3 + 3HNO_3 + (H_2SO_4) \rightarrow C_6H_7O_2(ONO_2)_3 + 3H_2O + (H_2SO_4)$$

(simplified formula).

The finished nitrocellulose should not be allowed to become acid in use or in storage, since this catalyzes its further decomposition. A stabilizer is therefore added that reacts with any trace of nitrous, nitric, or sulfuric acid that may be released because of the decomposition of the nitrocellulose and thus stop further decomposition.

See Dynamite.

NITROGEN

Nitrogen is a gas (boiling point: –196°C) that occupies approximately 80% of the volume of the air. Thus, it is not surprising that the production of nitrogen from the air is the predominant method of manufacture.

In the process for the liquefaction of air (Fig. 1), air is filtered to remove particulates and then compressed to 77 psi. An oxidation chamber converts traces of hydrocarbons into carbon dioxide and water. The air is then passed through a water separator that removes some of the water. A heat exchanger cools the sample down to very low temperatures, causing solid water and carbon dioxide to be separated from the main components.

Most of the nitrogen-oxygen mixture, now at –168°C and 72 psi, enters the bottom of a fractionating column (approximately 100 feet high) where an expansion valve at this point causes further cooling. The more volatile nitrogen rises to the top of the column as a gas since nitrogen (boiling point –196°C, 77 K) has a lower boiling point than oxygen (boiling point –183°C, 90 K), and the column at 83 K is able to separate the two.

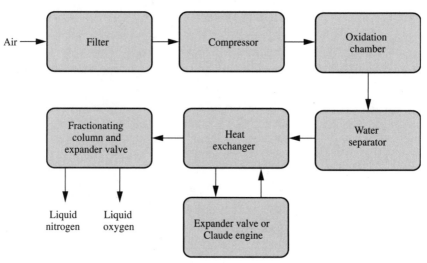

FIGURE 1   Manufacture of nitrogen by the liquefaction of air.

After being cooled in the heat exchanger, a small amount of nitrogen-oxygen mixture is fed to the main expander valve and the extremely cold gas is recycled into the heat exchanger to keep the system cold. Some argon remains in the oxygen fraction and this mixture can he sold as 90 to 95% purity oxygen. If purer oxygen is required, a more elaborate fractionating column with a greater number of plates gives an oxygen-argon separation. Oxygen can be obtained in 99.5% purity in this fashion. Not only argon, but also other rare gases, (neon, krypton, and xenon), can also be obtained in separations. Helium is *not* obtained from liquefaction of air. It occurs in much greater concentrations (2%) in natural gas wells and is isolated in the petroleum refinery.

Additional processes for the separation of pure nitrogen involve the use of cryogenics (Fig. 2), membranes (Fig. 3), or pressure-swing adsorption systems (Fig. 4). The use of any one of these processes either as a single adjunct or as a multiadjunct to the nitrogen process is known.

By far the largest use of nitrogen is in ammonia synthesis, and the fastest growing use of nitrogen is in enhanced oil recovery (EOR), where it maintains pressure in oil fields so that a vacuum is not formed underground when natural gas and oil are pumped out. It is competing with carbon dioxide in this application. Other uses include blanket atmospheres, food preservation, aerospace, cryogenics, metals processing, and electronic manufacturing.

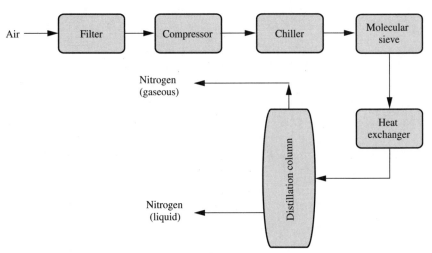

FIGURE 2 Nitrogen separation by cryogenics.

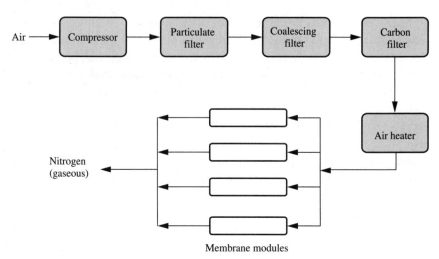

FIGURE 3   Nitrogen separation by the use of a membrane.

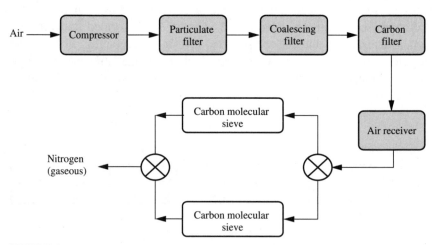

FIGURE 4   Nitrogen separation by pressure swing adsorption.

NITROGLYCERIN

Nitroglycerin ($CH_2ONO_2OCHNO_2CH_2ONO_2$, freezing point: 13°C) is a liquid similar in appearance to the original glycerol. However, the nitro compound is sensitive to percussion but is somewhat less sensitive in the solid phase. The solid tends to explode incompletely, so frozen nitroglycerin must always be thawed before using. To make nitroglycerin safer and easier to handle, it is usually manufactured into dynamite.

Nitroglycerin is manufactured by slowly (and cautiously!) adding glycerol ($CH_2OHCHOHCH_2OH$) of high purity (99.9%+) to a mixture having the approximate composition: sulfuric acid (H_2SO_4) 59.5%, nitric acid (HNO_3) 40%, and water (H_2O) 0.5%. Nitration is accomplished in agitated nitrators equipped with steel cooling coils carrying brine at 5°C to maintain the temperature below 10°C (Fig. 1).

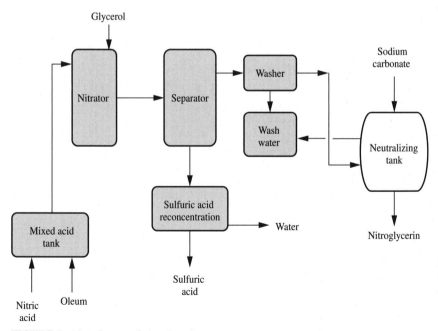

FIGURE 1   Manufacture of nitroglycerin.

After nitration, the mixture of nitroglycerin and spent acid is allowed to flow through a trough (a trough is easier to clean completely than a pipe) into separating and settling tanks some distance from the nitrator. The nitroglycerin is separated from the acid and sent to the wash tank, where it is washed twice with warm water and with a 2% sodium carbonate solution to ensure the complete removal of any remaining acid. Additional washes with warm water are continued until no trace of sodium carbonate (alkalinity) is evident.

See Dynamite.

NITROUS OXIDE

Nitrous oxide is manufactured by heating very pure ammonium nitrate to 200 to 260°C in aluminum retorts,

$$NH_4NO_3 \rightarrow N_2O + 2H_2$$

It is purified by treatment with caustic to remove nitric acid and with dichromate to remove nitric oxide.

See Ammonium Nitrate.

NONENE

Nonene (*n*-nonene; boiling point: 146°C, density: 0.730) was originally made by the trimerization of propylene ($CH_3CH=CH_2$) to give a branched nonene; this product now has limited use for detergents because of non-biodegradability.

Cracking and dehydrogenation of *n*-paraffins is now the preferred method, giving very linear chains.

$$C_9H_{20} \rightarrow C_9H_{18} + H_2$$

With wax consisting of linear paraffins, an olefin product containing as much as 90% linear alpha olefins can be prepared.

Nonene is used in the manufacture of nonylphenol and ethoxylated nonylphenol nonionic surfactants. It is also used in the oxo process to make *iso*-decyl alcohol for esters as plasticizers.

See Oxo Reaction.

NOVOCAINE

Novocaine (procaine hydrochloride) is a local anesthetic that is considered to be less toxic than cocaine, and does not have the danger of habituation. It is used frequently in conjunction with a vasoconstrictor such as epinephrine to secure a prolonged anesthetic action.

Alkylating ethylenechlorohydrin with diethylamine, which is condensed with *p*-nitrobenzoyl chloride and reduced with tin and hydrochloric acid to obtain procaine, produces novocaine.

NYLON

See **Caprolactam and Polyamides.**

OCHER

Ocher is a naturally occurring pigment consisting of clay colored with 10 to 30% ferric hydroxide [$Fe(OH)_3$]. It must be ground and levigated.

Ocher is a weak tinting agent and is often replaced by synthetic hydrated yellow iron oxides for brighter color and better uniformity.

ISO-OCTANE

Iso-octane (2,2,4-trimethylpentane; boiling point: 99.2°C, density: 0.6918, flash point: −12°C) is produced from *iso*-butylene by catalytic dimerization followed by hydrogenation (Fig. 1).

$$2(CH_3)_2C{=}CH_2 \rightarrow (CH_3)_3CCH_2CH(CH_3)_2$$

Fractionation is required so that the product can to meet the desired specifications.

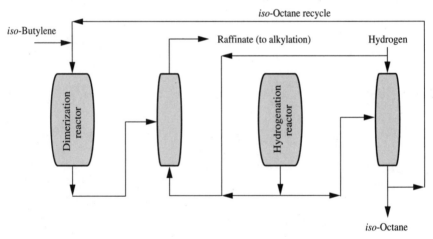

FIGURE 1   Manufacture of *iso*-octane.

OXYGEN

Oxygen is colorless, odorless, and tasteless as a gas, but it is slightly blue in the liquid state. Up to 99.995% purity is available commercially.

The manufacture of oxygen is described along with that of nitrogen (Fig. 1) during the liquefaction of air. Pressure swing adsorption (Fig. 2) is also used to generate pure oxygen.

Oxygen is used for primary metals manufacturing, chemicals manufacturing, oxidation processes, and partial oxidation processes. The steel industry prefers to use pure oxygen rather than air in processing iron. The oxygen reacts with elemental carbon to form carbon monoxide, which is processed with iron oxide so that carbon is incorporated into the iron metal, making it much lower melting and more pliable (fusible pig iron). The following equations summarize some of this chemistry.

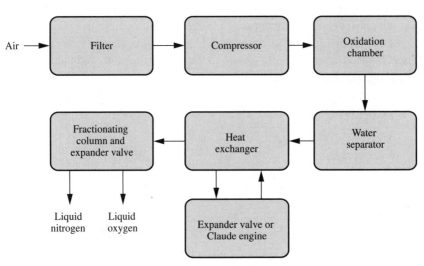

FIGURE 1 Manufacture of oxygen by the liquefaction of air.

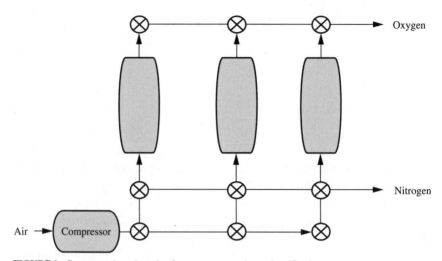

FIGURE 1   Pressure swing adsorption for oxygen generation and purification.

$$2C + O_2 \rightarrow 2CO$$

$$Fe_2O_3 + 3C \rightarrow 2Fe + 3CO_2$$
$$2CO \rightarrow C(\text{in Fe}) + CO_2$$

Common pig iron contains 4.3% carbon and melts at 1130°C, whereas pure iron has a melting point of 1539°C.

In other oxygen applications, metal fabrication involves cutting and welding with an oxygen-acetylene torch. Chemical manufacture use includes the formation of ethylene oxide, acrylic acid, propylene oxide, and vinyl acetate. Miscellaneous uses include sewage treatment, aeration, pulp and paper bleaching, and missile fuel.

See Nitrogen.

PAINTS

Liquid paint is a dispersion of a finely divided pigment in a liquid (the vehicle) composed of a resin or binder and a volatile solvent (Fig. 1). The pigment, although usually an inorganic substance, may also be a pure, insoluble organic dye known as a *toner*, or an organic dye precipitated on an inorganic carrier such as aluminum hydroxide, barium sulfate, or clay, thus constituting a *lake*.

The solid particles in the paint reflect many of the destructive light rays, and thus help to prolong the life of the paint. In general, pigments should be opaque to ensure good covering power and chemically inert to secure stability, hence long life.

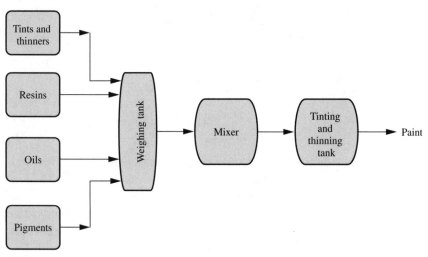

FIGURE 1   Manufacture of paint.

N-PARAFFINS

The production of n-paraffins, especially C_{10}–C_{14}, involves the use of zeolites to separate straight chain compounds from the kerosene fraction of petroleum.

The main use of n-paraffins is in the production of linear alkylbenzenes for the detergent industry and the production of linear alcohols, solvents, and chlorinated paraffins.

See Synthesis Gas.

PARALDEHYDE

Paraldehyde (2,4,6-trimethyl-1,3-5-trioxane, boiling point 125.4°C), a cyclic trimer of acetaldehyde, is formed when a mineral acid, such as sulfuric acid, phosphoric acid, or hydrochloric acid, is added to acetaldehyde. Paraldehyde can also be formed continuously by feeding liquid acetaldehyde at 15 to 20°C over an acid ion-exchange resin.

Depolymerization of paraldehyde occurs in the presence of acid catalysts, and, after neutralization with sodium acetate, acetaldehyde and paraldehyde are recovered by distillation.

See *Acetaldehyde.*

PENICILLIN

Penicillin is practically nontoxic (except to certain species of mold) and was the first antibiotic to be produced for widespread use.

The current viable process for the production of penicillin is large-scale fermentation. Large tanks of 5000 to 30,000-gal capacity are used. The penicillin is separated by solvent extraction. The mold grows best at 23 to 25°C, pH 4.5 to 5.0. The fermentation broth is made from corn steep liquor with lactose and inorganic materials added. Sterile air permits growth of the mold over a 50- to 90-hour period.

PENTAERYTHRITOL

Pentaerythritol (melting point: 261°C, density: 1.396, flash point: 260°C) belongs to the class of polyhydric alcohols, or polyols, that contain three or more methylene hydroxyl (CH_2OH) functional groups and have the general formula $R(CH_2OH)_n$, where $n = 0.3$ and R is an alkyl group or CCH_2OH group.

The most important polyhydric alcohols (Fig. 1) are white solids, ranging from crystalline pentaerythritol to the waxy trimethylol alkyls. The trihydric alcohols are very soluble in water, as is ditrimethylol propane. Pentaerythritol is moderately soluble and dipentaerythritol anti-tripentaerythritol are less soluble.

Pentaerythritol is manufactured by reaction of formaldehyde and acetaldehyde in the presence of a basic catalyst, generally an alkali or alkaline-earth hydroxide.

In the process (Fig. 2), the main concern in mixing is to avoid loss of temperature control in this exothermic reaction, which can lead to excessive by-product formation and/or reduced yields of pentaerythritol. The reaction time depends on the reaction temperature and may vary from about 0.5 to 4 hours at final temperatures of about 65 and 35°C, respectively. The reactor product, neutralized with acetic or formic acid, is then stripped of excess formaldehyde and water to produce a highly concentrated solution of pentaerythritol reaction products. This is then cooled under carefully controlled crystallization conditions so that the crystals can be readily separated from the liquors by subsequent filtration.

Staged reactions, where only part of the initial reactants is added, either to consecutive reactors or with a time lag to the same reactor, may be used to reduce dipentaerythritol content. This technique increases the effective formaldehyde-to-acetaldehyde mole ratio, maintaining the original stoichiometric one. It also permits easier thermal control of the reaction. Both batch and continuous reaction systems are used.

Dipentaerythritol and tripentaerythritol are obtained as by-products of the pentaerythritol process and may be further purified by fractional crystallization or extraction.

CH₂OH

HOCH₂ — C — CH₂OH   HOCH₂ — C — CH₂ — O — CH₂ — C — CH₂OH

CH₂OH

(1)   (2)
pentaerythritol,   dipentaerythritol
tetramethylolmethane

CH₂OH   CH₂OH   CH₂OH

HOCH₂ — C — H₂C — O — CH₂ — C — CH₂ — O — CH₂ — C — CH₂OH

CH₂OH   CH₂OH   CH₂OH

(3)
tripentaerythritol

CH₂OH   CH₂OH

H₃C — C — CH₂OH   H₃C — CH₂ — C — CH₂OH

CH₂OH   CH₂OH

(4)   (5)
trimethylolethane   trimethylolpropane

CH₂OH   CH₂OH

H₃C — CH₂ — C — CH₂ — O — CH₂ — C — CH₂ — CH₃

CH₂OH   CH₂OH

(6)
ditrimethylolpropane

Systematic names:

1. 2,2-bis(hydroxymethyl)1,3-propanediol
2. 2,2-oxybismethylene-bis(2-hydroxymethyl)-1.3-propanediol
3. 2,2-bis(3-hydroxy)-2,2-bis(hydroxymethyl)propoxy]methyl-1,3-propanediol
4. 2-hydroxymethyl-2-methyl-1,3-propanediol
5. 2-ethyl-2-hydroxymethyl-1,3-propanediol
6. 2,2-oxybismethylene-bis(2-ethyl)-1,3-propanediol

FIGURE 1 Formulae and names of the polyhydric alcohols.

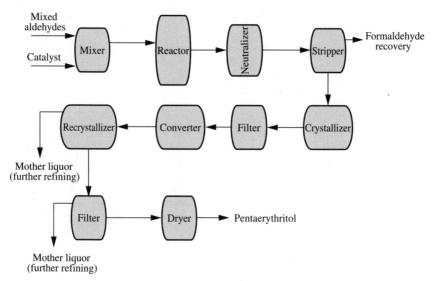

FIGURE 1   Manufacture of pentaerythritol.

PERACETIC ACID

There are two commercial processes for the production of peracetic acid: (1) Low-temperature oxidation of acetaldehyde in the presence of metal salts, ultraviolet irradiation, or treatment with ozone yields acetaldehyde monoperacetate, which can be decomposed to peracetic acid and acetaldehyde.

$$CH_3CH=O + [2O] \rightarrow CH_3CO-O-OH$$

(2) Peracetic acid can also be formed directly by liquid-phase oxidation at 5 to 50°C with a cobalt salt catalyst. Nitric acid oxidation of acetaldehyde yields glyoxal and the oxidation of p-xylene to terephthalic acid and of ethanol to acetic acid is activated by acetaldehyde.

PERCHLOROETHYLENE

There are three processes used in the making of perchloroethylene (tetra-chloroethylene; melting point: $-19°C$, boiling point: $121°C$, density: 1.6227), but the majority is made from ethylene dichloride.

$$C_2H_4Cl_2 + 5Cl_2 \rightarrow C_2H_2Cl_4 + C_2HCl_5 + 5HCl$$
$$C_2H_4Cl_2 + \rightarrow C_2Cl_4 + 2HCl$$

Perchloroethylene and trichloroethylene are produced in a single-stage oxychlorination process from ethylene dichloride and chlorine.

$$4C_2H_4Cl_2 + 6Cl_2 + 7O_2 \rightarrow 4C_2HCl_3 + 4C_2Cl_4 + 14H_2O$$

Chlorination of hydrocarbons, such as propane, and acetylene with chlorine also produces perchloroethylene via trichloroethylene.

1,1,1-trichloroethylene is usually made in the same apparatus or as a co-product. Both chlorination and oxychlorination are used to supply the reagents needed. The reactions follow the same pattern as those for ethane and methane chlorination. Temperatures, pressures, and reagent ratios are somewhat different, however.

The main use of perchloroethylene is in dry cleaning and textile pro-cessing; other uses are as a chemical intermediate, in industrial metal cleaning (vapor and cold degreasing), in adhesives, in aerosols, and in elec-tronics.

PETN

PETN [pentaerythritol tetranitrate, $C(CH_2ONO_2)_4$] is an extremely sensitive high explosive. When used as a booster explosive, a bursting charge, or a plastic demolition charge, it is desensitized by mixture with trinitrotoluene or by the addition of wax.

PETN is made by the nitration of pentaerythritol with strong (96%) nitric acid at about 50°C.

$$C(CH_2OH)_4 + 4HNO_3 \rightarrow C(CH_2ONO_2)_4 + 4H_2O$$

PETN is used in making detonating fuses and commercial blasting caps, and also has a medicinal use.

PETROCHEMICALS

Petrochemicals are relatively pure, identifiable substances derived from petroleum or natural gas (Fig. 1). Thus, ammonia and synthetic rubber made from natural gas components can be classed as petrochemical compounds. Among the most important petrochemicals manufactured include:

Acetic acid	Ethylene dichloride	Phenol
Acetone	Ethylene glycol	Polyethylene
Acrylonitrile	Ethylene oxide	Polypropylene
Benzene	Formaldehyde	Polyvinyl chloride
Cumene	Isopropyl alcohol	Styrene
Cyclohexane	Maleic anhydride	Toluene
Ethylbenzene	Methanol	Vinyl chloride
Ethylene	Phthalic anhydride	Xylenes

Refinery operations such as distillation, extraction, and various separation operations, and the chemical unit processes, such as alkylation, dehydrogenation, hydrogenation, and isomerization, are essentially identical to those operations used in the manufacture of chemicals from other sources.

Most processes for separating individual species from petroleum involve use of refined engineering methods, with distillation and selective adsorption the most important. Once separated, however, most materials then undergo chemical conversion into more desirable products. Alkylation involving propene and butenes yields C_6 to C_8 hydrocarbons for high-octane gasoline. Propylene becomes polypropylene, propylamine, or propylene glycol and ethers.

The most basic raw petrochemical materials are liquefied petroleum gas, natural gas, gas from cracking operations, liquid distillate (C_4 to C_6), distillate from special cracking processes, and selected or isomerized cyclic fractions for aromatics. Mixtures are usually separated into their components at the petroleum refineries, then chemically converted into reactive precursors before being converted into salable chemicals within the plant.

The lower members of the paraffin and olefin series have been the preferred and most economical sources of organic raw material for conversion, so figures and tables are shown concerning the derivations from methane (Fig. 2), ethylene (Fig. 3), propylene and butylene (Fig. 4), and ring-containing (cyclic) chemicals (Fig. 5).

Starting material	Process	Product
Petroleum	Distillation	Light ends
		Methane
		Ethane
		Propane
		Butane
	Catalytic cracking	Ethylene
		Propylene
		Butylenes
		Higher olefins
	Coking	Ethylene
		Propylene
		Butylenes
		Higher olefins
Natural gas	Refining	Methane
		Ethane
		Propane
		Butane

FIGURE 1 Production of starting materials for a petrochemical refinery.

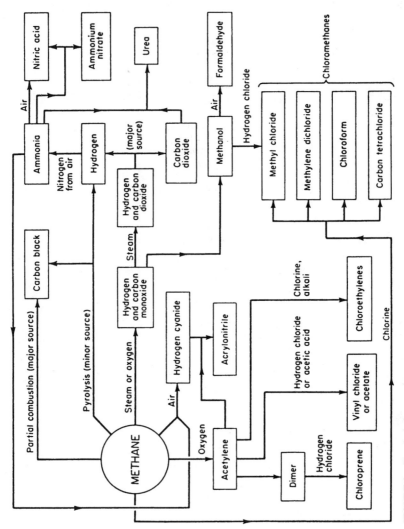

FIGURE 2 Chemicals from methane.

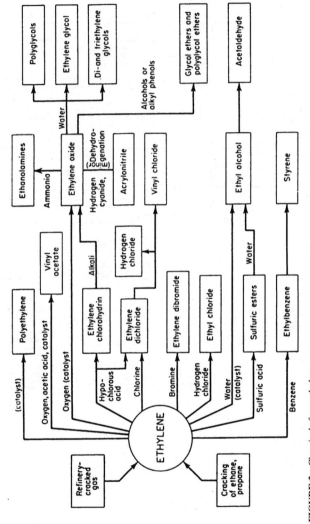

FIGURE 3 Chemicals from ethylene.

2.385

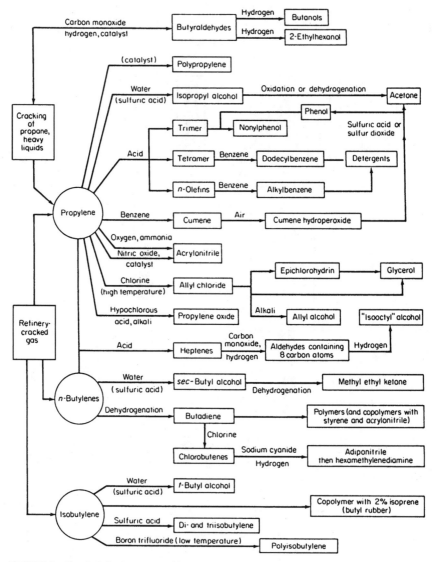

FIGURE 4   Chemicals from propene and butene(s).

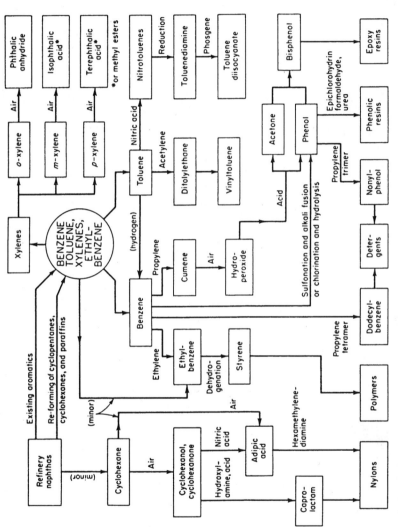

FIGURE 5 Cyclic chemicals.

PHENOBARBITAL

Phenobarbital (5-ethyl-5-phenylbarbituric acid) possesses specific usefulness in epilepsy.

Phenobarbital, like other barbituric acid derivatives, is manufactured from phenylethylmalonic diethyl ester, which is condensed with urea to form the product.

PHENOL

Phenol (hydroxybenzene; freezing point: 40.9°C, boiling point: 181.8°C, density: 1.0722; flash point: 79°C) at room temperature is a white, crystalline mass. Phenol gradually turns pink if it contains impurities or is exposed to heat or light. It has a distinctive sweet, tarry odor, and burning taste. Phenol has limited solubility in water between 0 and 65°C. Above 65.3°C, phenol and water are miscible in all proportions. Phenol is very soluble in alcohol, benzene, chloroform, ether, and partially disassociated organics in general, but it is less soluble in paraffinic hydrocarbons.

Phenol has been made, over the years, by a variety of processes. Historically, an important method was the sulfonation of benzene followed by desulfonation with caustic soda:

$$C_6H_6 + H_2SO_4 \rightarrow C_6H_5SO_3H + H_2O$$
$$C_6H_5SO_3H + 2NaOH \rightarrow C_6H_5OH + Na_2SO_4$$

This route to phenol is no longer used.

The principal process in use is the peroxidation of cumene (*iso*-propyl benzene) at 130°C in the presence of air and a catalyst followed by decomposition of the peroxide at 55 to 65°C in the presence of sulfuric acid.

$$C_6H_5CH(CH_3)_2 \rightarrow C_6H_5C(CH_3)_2OOH$$
$$C_6H_5C(CH_3)_2OOH \rightarrow C_6H_5OH + CH_3COCH_3$$

In the *cumene process* (Fig. 1), cumene is oxidized to form cumene hydroperoxide that is then concentrated and cleaved to produce phenol and acetone. By-products of the oxidation reaction are acetophenone and dimethyl benzyl alcohol, which is dehydrated in the cleavage reaction to produce alpha-methylstyrene.

The *toluene-benzoic acid process* involves three chemical reactions: (1) oxidation of toluene to form benzoic acid, (2) oxidation of benzoic acid to form phenyl benzoate, and (3) hydrolysis of phenyl benzoate to form phenol.

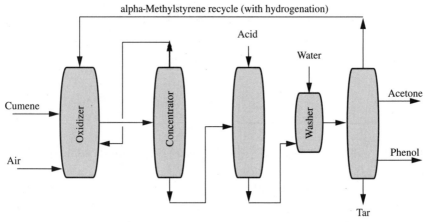

FIGURE 1   Manufacture of phenol from cumene.

$$2C_6H_5CH_3 + 3O_2 \rightarrow 2C_6H_5CO_2H + 2H_2O$$
$$4C_6H_5CO_2H + O_2 \rightarrow 2C_6H_5CO_2C_6H_5 + 2H_2O + 2CO_2$$
$$C_6H_5CO_2C_6H_5 + 2OH^- \rightarrow 2C_6H_5OH + CO_2$$

A typical process (Fig. 2) consists of two continuous steps. I n the first step the oxidation of toluene to benzoic acid is achieved with air and cobalt salt catalyst at a temperature between 121 and 177°C (206 kPa gauge), and the catalyst concentration is between 0.1 and 0.3%. The reactor effluent is distilled, and the purified benzoic acid is collected. In the second processing step, the benzoic acid is oxidized to phenyl benzoate in the presence of air and a catalyst mixture of copper and magnesium salts (234°C, 147 kPa gauge). The phenyl benzoate is then hydrolyzed with steam in the second reactor to yield phenol and carbon dioxide (200°C and atmospheric pressure).

Other processes include the production of *phenol from cyclohexene* in which phenol is produced from cyclohexene (benzene is partially hydrogenated to cyclohexene in the presence of water and a ruthenium-containing catalyst). The cyclohexene is then reacted with water to form cyclohexanol or oxygen to form cyclohexanone, and the cyclohexanol or cyclohexanone is dehydrogenated to phenol.

In the *benzene sulfonation process*, benzene reacts with concentrated sulfuric acid to form benzenesulfonic acid at about 150°C. The benzenesulfonic acid is neutralized with sodium sulfate to produce sodium benzenesulfonate, which is then fused with caustic soda to yield sodium phenate, which, after acidification with sulfur dioxide and a small amount of sulfuric acid, releases phenol from the sodium salt.

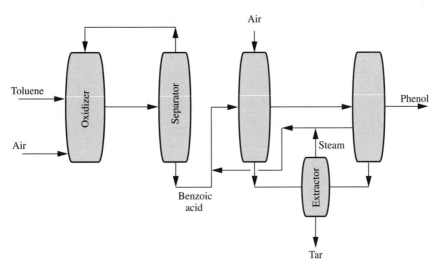

FIGURE 2   Manufacture of phenol from toluene.

In the *chlorobenzene* process, benzene is chlorinated at 38 to 60°C in the presence of ferric chloride (FeCl₃) catalyst. The chlorobenzene is hydrolyzed with caustic soda at 400°C and 11,000 psi (2.56 kPa) to form sodium phenate. The impure sodium phenate reacts with hydrochloric acid to release the phenol from the sodium salt.

In the *benzene oxychlorination process*, also known as the Raschig process, benzene is oxychlorinated with hydrogen chloride and air in the presence of iron and copper chloride catalyst to form chlorobenzene. The reaction occurs at 200 to 260°C and atmospheric pressure. The chlorobenzene is hydrolyzed at 480°C in the presence of a suitable catalyst to produce phenol and chloride.

Phenol is used in the manufacture of formaldehyde resins, bisphenol A, caprolactam, aniline, xylenols, and alkylphenols. Phenol-formaldehyde polymers (phenolic resins) have a primary use as the adhesive in plywood formulations. The use of phenol in detergent synthesis to make alkylphenols is an important aspect of phenol utility.

See Cumene, Cyclohexane, Toluene.

PHENOLIC RESINS

Phenolic resins (phenol-formaldehyde polymers), copolymers of phenol and formaldehyde, were the first fully synthetic polymers made. They were discovered in 1910 by Leo Baekeland and given the trade name Bakelite®.

Two processes, both involving step growth polymerization, are used for the manufacture of phenolic resins.

A *one-stage* resin may be obtained by using an alkaline catalyst and excess formaldehyde to form linear, low-molecular-weight resol resins. Acidification and further heating causes the curing process to give a highly cross-linked thermoset polymer. The o- and p-methylolphenols are more reactive toward formaldehyde than the original phenol and rapidly undergo further reaction to give di- and trimethylol derivatives. The methylolphenols will react to form di- and trinuclear phenols at still-free ortho and para positions. The final structure of the product involves a high degree of branching. Most linkages between aromatic rings are methylene (CH_2) groups, though some ether ($CH_2 \cdot OCH_2$) linkages are present.

The second process (Fig. 1) uses an acid catalyst and excess phenol to give a linear polymer (*novolac*) that has no free methylol groups for cross-linking. In a separate second part of this two-stage process, a cross-linking agent is added and further reaction occurs. In many instances, hexamethylenetetramine is used, which decomposes to formaldehyde and ammonia.

Other modifications in making phenolic polymers are the incorporation of cresols or resorcinol as the phenol (Fig. 1) and acetaldehyde or furfural as the phenol.

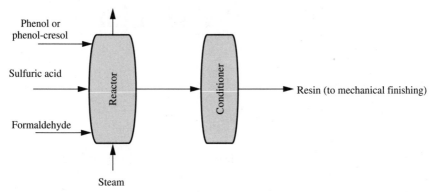

FIGURE 1   Manufacture of phenolic resins.

PHENOLPHTHALEIN

Phenolphthalein, in addition to its use an indicator in analytical chemistry, is a widely used cathartic agent.

Phenolphthalein is manufactured by adding melted phenol (10 parts) to a cooled solution of phthalic anhydride (5 parts) in concentrated sulfuric acid (4 parts) and heating the mixture 10 to 12 hours at 120°C. The hot condensation product is poured into boiling water and boiled with successive changes of hot water. The condensate is then dissolved in warm, dilute caustic soda and precipitated with acetic acid. It may be purified by crystallization from absolute alcohol after treatment with, and being filtered through activated carbon.

PHENOTHIAZINES

Phenothiazines are tranquilizing drugs in which the basic structure consists of two benzene rings fused to a central six-membered ring containing a sulfur and a nitrogen. They are sometimes administered as the hydrochloride salt by quaternarization of the side chain nitrogen.

The production of phenothiazines involves heating the appropriate meta-substituted diphenylamine with sulfur and an iodine catalyst to close the ring. Treatment with the strong base sodium amide gives the anion on the ring nitrogen, which then displaces the chlorine of the appropriate second reactant.

There are three main therapeutic applications of the phenothiazine drugs: (1) they have an antiemetic effect (stop vomiting); (2) they are used with anesthetics, potent analgesics (pain relievers), and sedatives to permit their use in smaller doses; and (3) they are used most widely to relieve anxiety and tension in various severe mental and emotional disorders.

PHENYLETHYL ALCOHOL

Phenylethyl alcohol has a roselike odor and occurs in the volatile oils of rose, orange flowers, and others. It is an oily liquid and is much used in perfume formulation.

Phenylethyl alcohol can be made by a number of procedures; the Grignard reaction is used generally:

$$C_6H_5Br + Mg \rightarrow C_6H_5MgBr$$

$$C_6H_5MgBr + CH_2CH_2O \rightarrow C_6H_5CH_2CH_2OMgBr$$
$$C_6H_5CH_2CH_2OMgBr + H^+ \rightarrow C_6H_5CH_2CH_2OH$$

However. the Friedel-Crafts reaction (Fig. 1) is also employed to manufacture this particular chemical.

$$C_6H_6 + CH_2CH_2O \, (+ AlCl_3) \rightarrow C_6H_5CH_2CH_2OH \, (+AlCl_3)$$

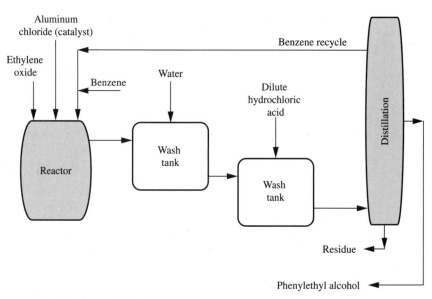

FIGURE 1   Manufacture of phenylethyl alcohol.

PHOSGENE

Phosgene is manufactured by reacting chlorine gas and carbon monoxide in the presence of activated carbon.

$$CO + Cl_2 \rightarrow O=CCl_2$$

Uses of phosgene include the manufacture of toluene diisocyanate, methylene diisocyanate, and polycarbonate resins.

PHOSPHORIC ACID

Phosphoric acid (H_3PO_4, melting point: 42°C), in the pure state, is a colorless solid. The usual laboratory concentration is 85% phosphoric acid, since a crystalline hydrate separates at 88% concentration.

Crude phosphoric acid is often black and contains dissolved metals and fluorine, and dissolved and colloidal organic compounds. Suspended solid impurities are usually removed by settling and solvent extraction (using a partially miscible solvent, such as *n*-butanol, *iso*-butanol, or *n*-heptanol), or solvent precipitation is used to remove the dissolved impurities. The phosphoric acid is extracted, and the impurities are left behind. Back-extraction with water recovers the purified phosphoric acid. Solvent precipitation uses a completely miscible solvent plus alkalis or ammonia to precipitate the impurities as phosphate salts. After filtration, the solvent is separated by distillation and recycled.

The majority of the phosphoric acid is made by the *wet process,* which involves reaction of calcium phosphate or the mixed calcium fluoride-calcium phosphate ores with sulfuric acid.

$$Ca_3(PO_4)_2 + 3H_2SO_4 \rightarrow 2H_3PO_4 + 3CaSO_4$$

$$CaF_2 \cdot Ca_3(PO_4)_2 + 10H_2SO_4 + 20H_2O \rightarrow 10(CaSO_4 \cdot 2H_2O) + 2HF + 6H_3PO_4$$

$$Ca_5F(PO_4)_3 + 5H_2SO_4 + 10H_2O \rightarrow 5(CaSO_4 \cdot 2H_2O) + HF + 3H_3PO_4$$

In the wet process (Fig. 1), the phosphate rock is ground and mixed with dilute phosphoric acid in a mill, after which it is transferred to a reactor and sulfuric acid is added. The reactors are heated to 75 to 80°C for 4 to 8 hours. Air cooling carries the hydrogen fluoride and silicon tetrafluoride side products to an adsorber that transforms them into fluorosilicate acid (H_2SiF_6). Filtration of the solid calcium sulfate ($CaSO_4 \cdot 2H_2O$, gypsum) gives a dilute phosphoric acid solution (28 to 35% P_2O_5 content) and evaporation of the water to a higher concentration (54% P_2O_5 content) is optional. The fluorosilicic acid (H_2SiF_6) is formed in the process by reac-

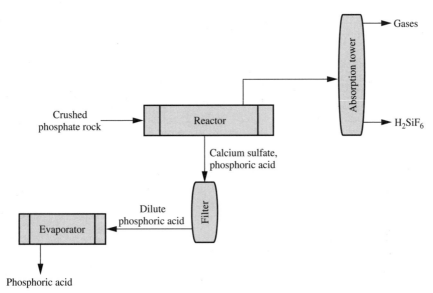

FIGURE 1   Manufacture of phosphoric acid by the wet process.

tion of the hydrogen fluoride with silica (SiO_2) that is present in most phosphate rock.

$$4HF + SiO_2 \rightarrow SiF_4 + 2H_2O$$
$$2HF + SiF_4 \rightarrow H_2SiF_6$$
$$3SiF_4 + 2H_2O \rightarrow 2H_2SiF_6 + SiO_2$$

The fluorosilicate acid can be used for fluoridation of drinking water, and fluorosilicate salts find use in ceramics, pesticides, wood preservatives, and concrete hardeners. The phosphoric acid can be purified by a solvent extraction process (Fig. 2).

Another wet process for producing phosphoric acid uses:

1. Hydrochloric acid to acidulate (in slight excess to prevent formation of monocalcium phosphate)

2. An organic solvent (C_3 or C_4 alcohols) to extract the phosphoric acid

3. Water to strip out the phosphoric acid (with a small amount of solvent and hydrochloric acid)

4. Concentration to remove the small amounts of solvent and hydrochloric acid and to yield a high-grade product

The *furnace process* is used only to make concentrated acid (75-85%) and pure product. It is very expensive because of the 2000°C temperature

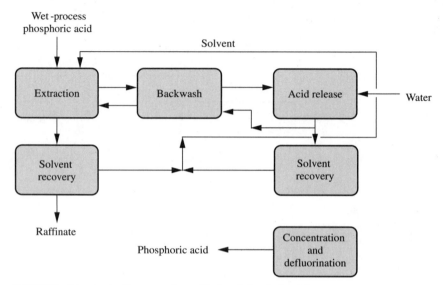

FIGURE 2   Solvent extraction process for purification of phosphoric acid.

required. In the furnace process, phosphate rock is heated with sand and coke to give elemental phosphorus, which is then oxidized and hydrated to phosphoric acid.

$$2Ca_3(PO_4)_2 + 6SiO_2 + 10C \rightarrow P_4 + 10CO + 6CaSiO_3$$
$$P_4 + SO_2 + 6H_2O \rightarrow 4H_3PO_4$$

See Phosphorus.

PHOSPHORUS

Yellow phosphorus (known also as *white phosphorus*) is produced by reducing phosphate rock (calcium phosphate or calcium fluorophosphate) with carbon in the presence of silica as flux; an electric arc furnace furnishes heat of reaction (Fig. 1).

$$2Ca_3(PO_4)_2 + 10C + 6SiO_2 \rightarrow P_4 + 6CaSiO_3 + 10CO$$

The silica is an essential raw material that serves as an acid and a flux. About 20 percent of the fluorine present in the phosphate rock is converted to SiF_4 and volatilized. In the presence of water vapor this reacts to give silica (SiO_2) and fluorosilicic acid (H_2SiF_6).

$$3SiF_4 + 2H_2O \rightarrow 2H_2SiF_6 + SiO_2$$

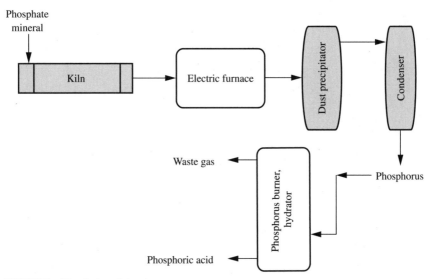

FIGURE 1   Manufacture of phosphorus.

Phosphorus is used for the manufacture of phosphoric acid and other chemicals, including phosphorus trichloride, phosphorus pentasulfide, and phosphorus pentoxide.

*See **Phosphoric Acid.***

PHTHALIC ACID

*See **Phthalic Anhydride**.*

PHTHALIC ANHYDRIDE

Phthalic anhydride (melting point: 131.6°C, boiling point: 295°C with sublimation) can be made from the reaction of *o*-xylene with air and also from
naphthalene (Fig. 1), which is isolated from coal tar and from petroleum.

$$CH_3C_6H_4CH_3 + [O] \rightarrow HOOCC_6H_4COOH$$
$$HOOCC_6H_4COOH \rightarrow OCC_6H_4COO + H_2O$$

The use of naphthalene as the source of phthalic anhydride diminished but
has become popular once again.

Phthalic anhydride is used for the manufacture of plasticizers, such as
dioctyl phthalate, unsaturated polyester resins, and alkyd resins. Phthalic
anhydride reacts with alcohols such as 2-ethylhexanol, and the products are
often liquids that, when added to plastics, impart flexibility without adversely
affecting the strength of the plastic. Most of these plasticizers are used for
poly(vinyl chloride) flexibility. Dioctyl phthalate is a common plasticizer.

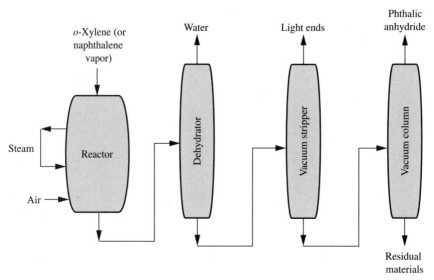

FIGURE 1 Manufacture of phthalic anhydride.

PHTHALOCYANINE BLUE

The phthalocyanine blues are particularly useful for nitrocellulose lacquers in low concentrations as a pigment highly resistant to alkali, acid, and color change.

As presently produced, they are highly stable pigments, resistant to crystallization in organic solvents, and essentially free from flocculation in coatings. Both the green and the blue dyes have high tinting power and are used in latex paints and in printing inks, as well as all types of interior and exterior coatings. They are prepared by reacting phthalic anhydride with a copper salt with or without ammonia.

PHTHALOCYANINE GREEN

The major green pigment is phthalocyanine green. It is a complex copper compound that is suitable for use in both solvent- and water-based paints.

PICRIC ACID

Picric acid (2,4,6-trinitrophenol, melting point: 122°C) is manufactured by the nitration of mixed phenolsulfonates with mixed acid. Mixed acid increases the yield of desired products.

The heavy metal salts of picric acid are dangerously sensitive, and its major use is for the manufacture of ammonium picrate (Explosive D).

PIPERAZINE CITRATE

Piperazine citrate is produced from piperazine that is first manufactured by the cyclization of ethylene dibromide with alcoholic ammonia at 100°C. The citrate is formed in aqueous solution and crystallized out.

Piperazine citrate is used as an anthelmintic in the treatment of infections caused by pinworms and roundworms. It is also employed by veterinarians against various worms infecting domestic animals, including chickens.

POLYACETALDEHYDE

Polyacetaldehyde, a rubbery polymer with an acetal poly(oxymethylene) structure, is an unstable solid that depolymerizes to acetaldehyde

Polyacetaldehyde is formed by cationic polymerization using boron trifluoride (BF_3) in liquid ethylene ($CH_2=CH_2$). At temperatures below $-75°C$ using anionic initiators, such as metal alkyls in a hydrocarbon solvent, a crystalline isotactic polymer is obtained. Molecular weights of the products fall into the range 800,000 to 3,000,000.

See Acetaldehyde.

POLYAMIDES

Polyamides are polymers that contain the amide group and are produced, by, for example, polymerization reaction of adipic acid and hexamethylenediamine (HMDA). Nylon 6,6 (so-called because each of the raw material chains contains six carbon atoms) was the first all-synthetic fiber made commercially.

Nylon 6 is the homopolymer of caprolactam, and Kevlar is an aromatic polyamide, poly-*p*-phenylene terephthalamide.

The reaction between adipic acid and hexamethylene diamine produces hexamethylenediammonium adipate, commonly called *nylon salt* (Fig. 1) Forming the salt assures the correctly balanced proportions. It is also necessary that the material to be polymerized contain very few impurities if high-quality fibers are to be made by a variety of routes (Fig. 2).

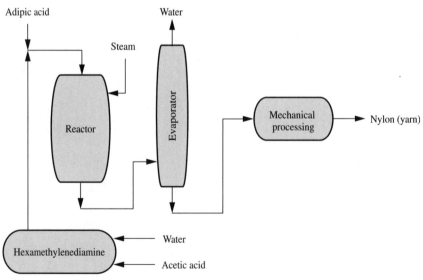

FIGURE 1   Manufacture of nylon.

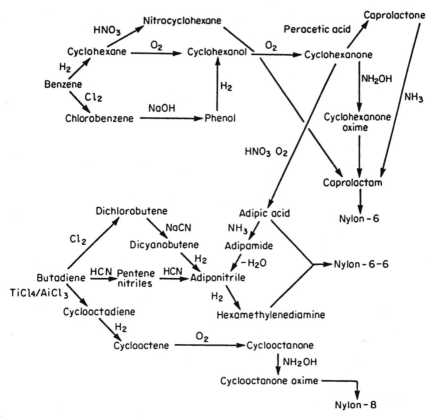

FIGURE 2 Various routes to nylon.

Like all other synthetic fibers that have become competitively popular, nylon in both the filament and staple form must have certain properties that are superior to natural fibers. It is stronger than *any* natural fiber and has a wet strength of 80 to 90 percent of its dry strength. Its good flexing tenacity makes it very desirable for women's hosiery, and it has good stretch recovery. Nylon's high tenacity has made it important in parachute fabrics and related nonapparel items. Nylon can be dyed by all acid and dispersed dyes. It has a low affinity for direct cotton, sulfur, and vat dyes.

Nylon 6, or nylon caprolactam, is a polymeric fiber derived from only one constituent, caprolactam, giving the polymer $[-(CH_2)_5CONH-]_n$. It has a lower melting point than nylon 6,6, but it is superior to it in resistance to light degradation, elastic recovery, fatigue resistance, and thermal stability.

*See **Caprolactam.***

POLYCARBONATES

Polycarbonate resins are a variety of polyester in which a derivative of carbonic acid is substituted for adipic, phthalic, or other acid and a diphenol is substituted for the more conventional glycols.

A number of methods for the manufacture of polycarbonates are available but the *melt process* and the *phosgenation process* are the most important.

POLYCHLORINATED BIPHENYLS

Polychlorinated biphenyls (PCBs, Arochlor®, Phenochlor®, and Clophen®) are chemically similar to the chlorinated insecticides.

Polychlorinated biphenyls are manufactured by chlorination of biphenyl. The conditions determine the degree of chlorination.

They were used to make more flexible and flame-retardant plastics and are still used as insulating fluids in electrical transformers since there is no substitute in the application. Their existence and persistence in the environment is well established, and they are classed as a pollutant.

POLYESTERS

The common polyester fibers are polymers of the ester formed from dimethyl terephthalate and ethylene glycol. The polymer melts at 270°C and has very high strength and elasticity. It is 3 times as strong as cellulose. It is also particularly resistant to hydrolysis (washing) and resists creasing. Hence it has been used in clothing, especially shirts (65/35 polyester/cotton most popular). Its excellent clarity has made it useful in photographic film and overhead transparencies. Poly(ethylene terephthalate) is known commonly by the trademarks Dacron®, Terylene®, and Fortrel® fibers and Mylar® for film.

Polyesters are manufactured in one of two ways: by direct reaction of a di-acid and a diol or by ester interchange of a di-ester and a diol. By far the most commercially useful polyester is poly(ethylene terephthalate).

Polymerization is a two-stage process (Fig. 1) in which the monomer is first prepared either by an ester interchange between dimethyl terephthalate and ethylene glycol, or by direct esterification of terephthalic acid.

$$p\text{--}H_3COOCC_6H_4COOCH_3 + 2HOCH_2CH_2OH \rightarrow$$
$$p\text{--}HOCH_2CH_2OOCC_6H_4COOCH_2CH_2OH + 2CH_3OH$$
ester monomer

or

$$p\text{--}HOOCC_6H_4COOH + 2HOCH_2CH_2OH \rightarrow$$
$$p\text{--}HOCH_2CH_2OOCC_6H_4COOCH_2CH_2OH + 2H_2O$$
ester monomer

The second stage is the polymerization of the monomer by heating at 260°C.

$$n(\text{ester monomer}) \rightarrow (n-1)HOCH_2CH_2OH +$$
$$H[OCH_2CH_2OOCC_6H_4CO]_nOCH_2CH_2OH$$
polyethylene terephthalate

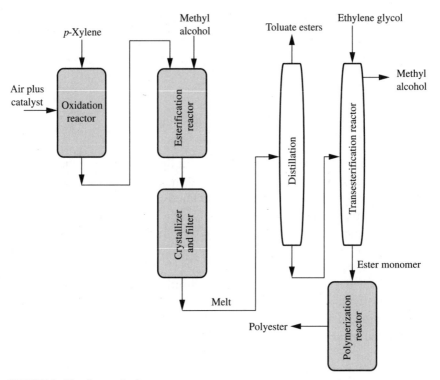

FIGURE 1   Manufacture of polyesters.

The polymer is extruded from the bottom of the polymerizer through a slot or holes on to the surface of a water-cooled drum. The ribbon is cut to chips and dried before melt spinning in a manner similar to that described for nylon. The filaments are stretched, with the application of heat, to about 3 to 6 times their original length.

Most useful polyesters have need for the strong, rigid aromatic ring in their structure since they lack the hydrogen bonding prevalent in polyamides.

POLYESTERS (UNSATURATED)

An unsaturated polyester resin consists of a linear polyester whose chain contains double bonds and an unsaturated monomer such as styrene that copolymerizes with the polyester to provide a cross-linked product.

The most common unsaturated polyester is manufactured by step growth polymerization of propylene glycol with phthalic and maleic anhydrides. Subsequent treatment with styrene and a peroxide catalyst leads to a solid, infusible thermosetting product.

Unsaturated polyesters are relatively brittle, and about 70 percent is used with fillers, of which glass fiber is easily the most popular. Glass fiber–reinforced polyester for small boat hulls consumes one-quarter of unsaturated polyesters.

POLYHYDRIC ALCOHOLS

See **Pentaerythritol.**

POLYIMIDES

Polyimides are compounds that contain two anhydride groups, which will react with primary amines or isocyanates to form polyimide polymers that are very stable linear polymers.

POLYSULFONES

Polysulfones are both aliphatic and aromatic polymers that are resistant to high temperatures and are very stable.

Typically, polysulfones are prepared by the reaction of disodium bisphenol A with 4,4'-dichlorodiphenylsulfone.

Their resistance to autoclave sterilization makes them useful for medical instruments and trays. Other uses are microwave cookware, coffee decanters, and corrosion-resistant piping.

POLYURETHANE FOAMS

Polyurethane foams are manufactured from a diisocyanate, such as toluene diisocyanate (TDI), and a low-molecular-weight polyether such as poly (propylene glycol).

In the process, a small amount of water is added to convert some isocyanate functionalities into carbon dioxide and amines. The degree of foaming can be controlled by the amount of water added.

One way of obtaining the more useful cross-linked polyurethanes is by using a trifunctional reagent. Thus either the toluene diisocyanate can react with a triol or the propylene oxide can be polymerized in the presence of a triol.

POTASSIUM CHLORATE

See **Sodium Chlorate.**

POTASSIUM COMPOUNDS

The industrial term *potash* usually refers to potassium carbonate (K_2CO_3) but also can be used in reference to potassium hydroxide (KOH), potassium chloride (KCl), potassium sulfate (K_2SO_4), potassium nitrate (KNO_3), or collectively to all potassium salts and to the oxide (K_2O). Another term, *caustic potash,* is used in reference to potassium hydroxide, and *muriate of potash* is commonly used in reference to potassium chloride.

Deposits of sylvinite (43% by weight potassium chloride and 57% by weight sodium chloride) account for large amounts of naturally occurring potassium. The potassium chloride can be separated by either fractional crystallization or flotation. Brine is also a valuable source of potassium chloride. A small amount of potassium sulfate is isolated from natural deposits, and potassium nitrate is manufactured by two processes.

$$NaNO_3 + KCl \rightarrow KNO_3 + NaCl$$
$$4KCl + 4HNO_3 + O_2 \rightarrow 2KNO_3 + Cl_2 + H_2O$$

Potassium hydroxide is made by electrolysis of potassium chloride solutions in cells that are exactly analogous to those for sodium hydroxide production.

$$2KCl + 2H_2O \rightarrow 2KOH + H_2 + Cl_2$$

POTASSIUM HYDROXIDE

Potassium hydroxide is produced by the electrolysis of potassium chloride solutions.

$$2KCl + 2H_2O \rightarrow 2KOH + H_2 + Cl_2$$

Potassium hydroxide is used for the production of potassium carbonate, tetrapotassium pyrophosphate and other potassium phosphates, liquid fertilizers, and soaps.

POTASSIUM NITRATE

Potassium nitrate (saltpeter, KNO_3) is manufactured in two ways: (1) by reacting nitric acid with potassium chloride with a chlorine by-product and (2) by reacting sodium nitrate with potassium chloride and crystallizing out the salt.

Potassium nitrate is used in the manufacture of fertilizers, explosives, ceramics, and heat-treating salts.

POTASSIUM PERCHLORATE

Potassium perchlorate is made by converting sodium chlorate into sodium perchlorate in steel electrolytic cells that have platinum anodes and operate at a temperature of 65°C. Filtered potassium chloride solution is added to the sodium perchlorate, precipitating potassium perchlorate crystals, which are centrifuged, washed, and dried. The mother liquor now contains sodium chloride, which can be used as a cell feed for sodium chlorate manufacture.

Potassium perchlorate may also be manufactured by electrolysis of potassium chloride solutions.

Potassium perchlorate is used in matches, pharmaceutical products, and pyrotechnic mixtures. Most mixtures of organic substances and chlorates are explosive, so use of any such mixtures should be limited to those expert in their use.

PRODUCER GAS

Producer gas is manufactured by passing air and steam through a bed of hot coal or coke at a temperature of 980 to 1540°C, depending on the fusion points of the fuel ash. The primary purpose of the steam (25 to 30 % of the weight of the coke) is to employ the exothermic energy from the reaction between carbon and oxygen to supply energy (heat) to the endothermic carbon-steam reaction.

$$C + O_2 \rightarrow CO_2$$

$$CO_2 + C \rightarrow 2CO$$

$$C + H_2O \rightarrow CO + H_2$$
$$CO + H_2O \rightarrow CO_2 + H_2$$

The initial reaction is the formation of carbon dioxide, and, as the gases progress up the bed, the carbon dioxide is reduced to carbon monoxide and the water vapor is partly decomposed to yield hydrogen, carbon monoxide, and carbon dioxide. Producer gas has about 15 percent of the heating value of natural gas.

Producer gas was once used for industrial heating, but its use is now diminished and it finds only occasional use in industrial operations.

PROPANE

Natural gas (predominantly methane, CH_4, with ethane, propane, and butane) is a fuel gas, and it is also an important chemical raw material for various syntheses. Natural gas can also be separated into its hydrocarbon constituents thereby producing feedstocks for petrochemical processes.

Propane ($CH_3CH_2CH_3$, melting point: $-189.7°C$, boiling point $-42.1°C$, density: 0.5853) is also a constituent of liquefied petroleum gas along with butane ($CH_3CH_2CH_2CH_3$), and propane can be used for the manufacture of aromatics by reforming (Fig. 1).

See *Liquefied Petroleum Gas, Natural Gas.*

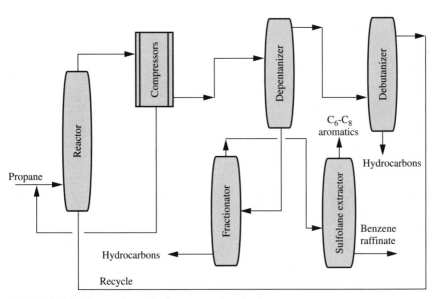

FIGURE 1   Manufacture of aromatics from propane by reforming.

PROPANOLOL HYDROCHLORIDE

This compound, sold under the trade name Inderal®, was the second most widely prescribed drug in the early 1980s. It is used as an antianginal and antihypertensive drug.

Propanolol hydrochloride is prepared from α-naphthol (1-naphthol) and epichlorohydrin. Subsequent treatment with isopropyl amine opens the epoxy ring to yield propanolol. Treatment with hydrochloric acid yields the hydrochloride.

Propanolol is a type of antihypertensive agent called a β-adrenergic blocking agent because it competes with epinephrine (adrenaline) and norepinephrine at their receptor sites and protects the heart against undue stimulation.

PROPARGYL ALCOHOL

Propargyl alcohol (2-propyn-1-ol, boiling point: 114°C) is a colorless volatile liquid with an unpleasant odor that is the only commercially available acetylenic primary alcohol. It is miscible with water and with many organic solvents. The commercial material is specified as 97% minimum purity, determined by gas chromatography or acetylation. Moisture is specified at 0.05% maximum (Karl-Fischer titration). Formaldehyde content is determined by bisulfite titration.

Propargyl alcohol is a by-product of butynediol manufacture. The original high-pressure butynediol processes gave about 5 percent of the by-product; lower-pressure processes give much less. Processes have been described that give much higher proportions of propargyl alcohol.

Although propargyl alcohol is stable, it has a low flash point, and vapor ignition and violent reactions can occur in the presence of contaminants, particularly at elevated temperatures. Heating in undiluted form with bases or strong acids should be avoided. Weak acids have been used to stabilize propargyl alcohol prior to distillation. Propargyl alcohol is a primary skin irritant and a severe eye irritant and is toxic by all means of ingestion; all necessary precautions must be taken to avoid contact with liquid or vapors.

Propargyl alcohol is a component of oil-well acidizing compositions, inhibiting the attack of mineral acids on steel, and is also employed in the pickling and plating of metals.

Propargyl alcohol is used as an intermediate in preparation of the miticide Omite® 2-(4'-tert-butylphenoxy)cyclohexyl 2-propynyl sulfite, of sulfadiazine, and of halogenated propargyl carbonate fungicides.

The chemical properties and uses of propargyl alcohol has three potentially reactive sites: (1) a primary hydroxyl group (i.e., CH_2OH), (2) a triple bond ($-C\equiv C-$), and (3) an acetylenic hydrogen ($-C\equiv CH$) that makes the alcohol an extremely versatile chemical intermediate. The hydroxyl group can be esterified with acid chlorides, anhydrides, or carboxylic acids, and it reacts with aldehydes or vinyl ethers in the presence of an acid catalyst to form acetals. At low temperatures, oxidation with chromic acid gives propynal or propynoic acid:

$$HC\equiv CCH_2OH \rightarrow HC\equiv CCH=O$$
$$HC\equiv CCH_2OH \rightarrow HC\equiv CCO_2H$$

Halogenating agents can be used to replace hydroxyl with chlorine or bromine. Phosphorus trihalides, especially in the presence of pyridine, are particularly suitable, and propargyl iodide is easily prepared from propargyl bromide by halogen exchange.

Hydrogenation gives allyl alcohol, its isomer propanal, or propanol, whereas with acidic mercuric salt catalysts, water adds to give acetol (hydroxyacetone):

$$HC\equiv CCH_2OH + H_2 \rightarrow CH_2=CHCH_2OH$$

$$HC\equiv CCH_2OH + H_2 \rightarrow CH_2=CHCH=O$$

$$HC\equiv CCH_2OH + H_2O \rightarrow CH_3COCH_2OH$$

In the presence of copper acetylide catalysts, propargyl alcohol and aldehydes give acetylenic glycols. When dialkylamines are also present, dialkylaminobutynols are formed.

*See **Butynediol.***

PROPENE

Propene (propylene, $CH_3CH=CH_2$, boiling point: $-47.7°C$, flash point: $-107.8°C$, ignition temperature: $497.2°C$) is a colorless, flammable gas with a slightly sweet aroma.

Like ethylene (q.v.), propylene can also be isolated from refinery gas but propylene (propene) is also manufactured by steam cracking of hydrocarbons as for ethylene and the best feedstocks are propane, naphtha, or gas oil (Fig. 1).

$$2CH_3CH_2CH_3 \rightarrow CH_3CH=CH_2 + CH_2=CH_2 + CH_4 + H_2$$

Much of the propylene is consumed by the petroleum refining industry for alkylation and polymerization to oligomers that are added to gasoline. A smaller amount is used for chemical manufacture.

Propene is used for the manufacture of a variety of chemicals (Fig. 2), including polypropylene, acrylic acid, *iso*-propyl alcohol, cumene, propy-

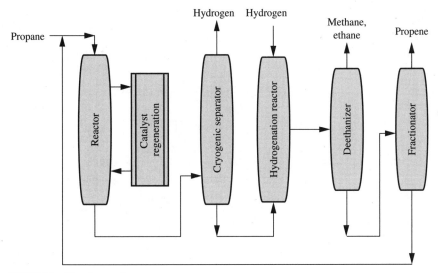

FIGURE 1   Manufacture of propane.

lene, and acrylonitrile. Propylene oxide is used in polyurethane plastic and foam. Cumene is made from propylene and benzene and is an important intermediate in the manufacture of phenol and acetone.

iso-propyl alcohol is made from propylene and is a common industrial solvent in coatings, chemical processes, pharmaceuticals, and household and personal products. Oxo chemicals are made by reacting propylene with synthesis gas (CO/H_2) to form C_4 alcohols. Small amounts of propylene are made into oligomers, where 3 to 5 propylene units are added to each other for use in the manufacture of soaps and detergents, in addition to being used as *polymer* gasoline.

See Iso-Propyl Alcohol.

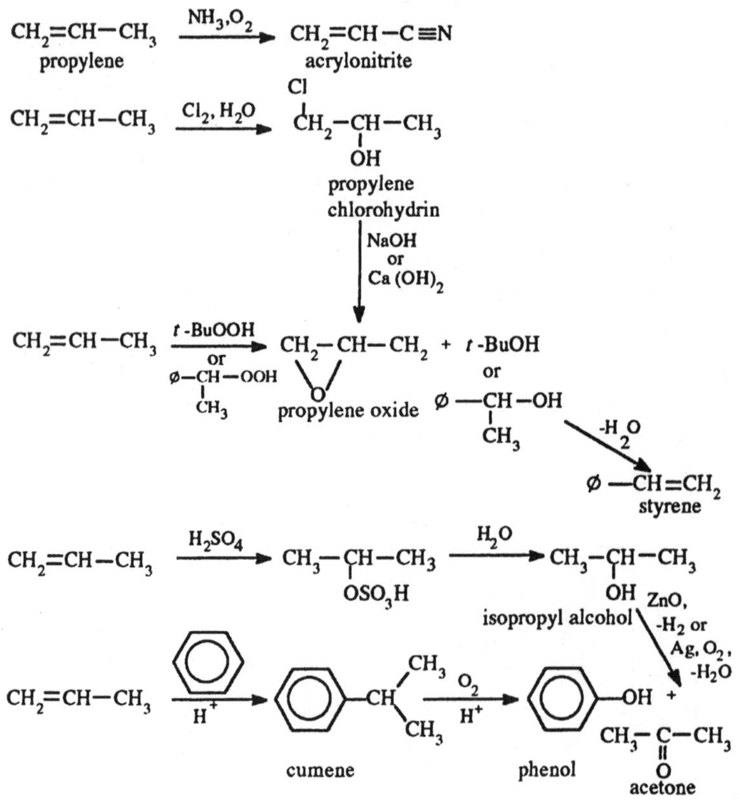

FIGURE 2   Manufacture of chemicals from propene.

ISO-PROPYL ALCOHOL

Iso-propyl alcohol (2-propanol, *iso*-propanol, rubbing alcohol; boiling point: 82.4°C, melting point: 89.5°C, density: 07855) is manufactured by the esterification/hydrolysis of propylene to *iso*-propyl alcohol. Unlike ethanol, for which the esterification/hydrolysis has been replaced by direct hydration, the direct process for *iso*-propyl alcohol is more difficult for crude propylene.

In the esterification process, only the propylene reacts and conditions can be maintained so that ethylene is inert.

$$CH_3CH=CH_2 + H_2SO_4 \rightarrow CH_3CH(OSO_3H)CH_3$$
$$CH_3CH(OSO_3H)CH_3 + H_2O \rightarrow CH_3CH(OH)CH_3 + H_2SO_4$$

The esterification step occurs with 85% sulfuric acid at 24 to 27°C, and dilution to 20% concentration is done in a separate tank. The *iso*-propyl alcohol is distilled from the dilute acid that is concentrated and returned to the esterification reactor. The *iso*-propyl alcohol is originally distilled as a 91% azeotrope with water. Absolute *iso*-propyl alcohol, boiling point 82.5°C, is obtained by distilling a tertiary azeotrope with *iso*-propyl ether. A 95% yield is realized.

Iso-propyl alcohol is used to produce acetone, pharmaceuticals, processing solvents, and coatings. Some of the chemicals derived from *iso*-propyl alcohol are *iso*-propyl ether (an industrial extraction solvent), *iso*-propyl acetate (a solvent for cellulose derivatives), *iso*-propyl myristate (an emollient, lubricant, and blending agent in cosmetics, inks, and plasticizers), *t*-butylperoxy *iso*-propyl carbonate (a polymerization catalyst and curing agent, and *iso*-propylamine and *diiso*-propylamine (low-boiling bases).

PROPYLENE GLYCOL

Propylene glycol (boiling point: 189°C, density: 1.0361) is produced by hydration of propylene oxide in a process similar to that for the production of ethylene glycol by hydration of ethylene oxide.

$$CH_3\overline{CHCH_2O} + H_2O \rightarrow CH_3CHOHCH_2OH$$

Unsaturated polyester resins account for the majority of the commercial use of propylene glycol. Other uses include food, cosmetics, pharmaceuticals, pet food, and tobacco humectants.

See Ethylene Glycol, Propylene Oxide.

PROPYLENE OXIDE

There are two important methods for the manufacture of propylene oxide. The older method involves chlorohydrin formation from the reaction of propylene with chlorine water (Fig. 1).

$$CH_3CH= CH_2 + Cl_2 + H_2O \rightarrow ClCH_2CH(OH)CH_3 + HCl$$
$$ClCH_2CH(OH)CH_3 + H_2O \rightarrow CH_3\overline{CHCH_2O}$$

The dilute cholorohydrin solution is mixed with a 10% slurry of lime to form the oxide, which is purified by distillation, boiling point 34°C. The yield is 90 percent.

A new variation of the chlorohydrin process uses *t*-butyl hypochlorite as the chlorinating agent. The waste brine solution can be converted back to chlorine and caustic by a special electrolytic cell to avoid the waste of chlorine.

The second manufacturing method for propylene oxide is via peroxidation of propylene (Halcon process). Oxygen is first used to oxidize *iso*-butane to *t*-butyl hydroperoxide (BHP) over a molybdenum naphthenate catalyst at 90°C and 450 psi (Fig. 2).

$$4CH_3CH(CH_3)CH_3 + 3O_2 \rightarrow 2CH_3C(CH_3)_2OOH + 2(CH_3)_3COH$$

The *t*-butyl hydroperoxide is then used to oxidize propylene to the oxide.

$$2CH_3C(CH_3)_2OOH + CH_3CH=CH_2 \rightarrow CH_3\overline{CHCH_2O} + (CH_3)_3COH$$

The *t*-butyl alcohol can be used to increase the octane number of unleaded gasoline or it can be made into methyl t-butyl ether (MTBE) for the same application. The alcohol can also be dehydrated to isobutylene, which in turn is used in alkylation to give highly branched dimers for addition to straight-run gasoline. An alternative reactant in this method is ethylbenzene hydroperoxide. This eventually forms phenylmethylcarbinol along with the propylene oxide, and the alcohol is dehydrated to styrene. Thus, the yield of the by-product can be varied depending on the demand for substances such as *t*-butyl alcohol or styrene.

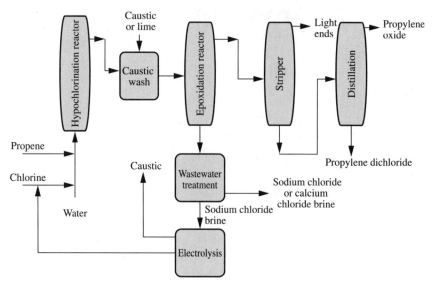

FIGURE 1   Manufacture of propylene oxide by the chlorohydrin process.

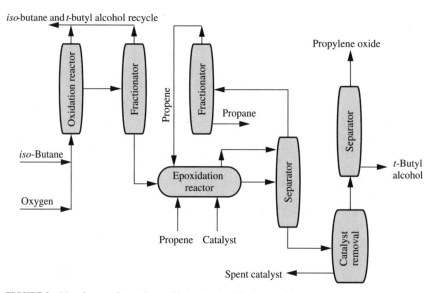

FIGURE 2   Manufacture of propylene oxide by the *t*-butyl hydroperoxide process.

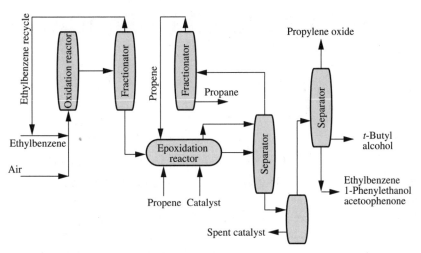

FIGURE 3   Manufacture of propylene oxide from ethylbenzene.

A third process involves the production of propylene oxide from ethylbenzene (Fig. 3); styrene is also produced.

Propylene oxide is used in the manufacture of propylene glycol, polypropylene glycol, dipropylene glycol, and glycol ethers. The polymerization of propylene oxide yields polypropylene glycol, which is actually a polyether, although it has terminal hydroxyl groups. These hydroxyl groups are reacted with an isocyanate such as toluene diisocyanate (TDI) to form the urethane linkages in the high-molecular-weight-polyurethanes.

PULP AND PAPER CHEMICALS

Cellulose is not only the most abundant organic substance available, it is a major component of woody plants and is constantly replaceable. Its conversion to paper products is the function of the pulp and paper industries, which manufacture thousands of useful items from cellulose.

Before paper can be made from wood, the cellulose fibers must be freed from the matrix of lignin that cements them together. The fibers may be separated by mechanical procedures or by solution of the lignin by various chemicals. The pulp thus formed has its fibers recemented together to form paper when suitable additives are used.

All processes used for pulping have the same goal, and that is the release the fibrous cellulose from its surrounding lignin while keeping the hemicelluloses and celluloses intact, thereby increasing the yield of useful fibers. The fibers thus obtained are naturally colored and must be bleached before they can be used for paper.

There are many processes and variations of basic processes that can be used for making pulp from wood. The major processes are: the *kraft process* (also known as the *sulfate process*), the groundwood and thermomechanical process, semichemical process, and the *sulfite process*. The kraft process remains dominant.

The kraft process (Fig. 1) is an alkaline process that is an outgrowth of the obsolete soda process that cooked with a strong (12%) solution of sodium hydroxide and sodium carbonate. The soda process gave low yields and worked well only with short-fibered hardwoods. The material added to the cooking liquor for the kraft process is sodium sulfate (Na_2SO_4, hence the alternative name, the sulfate process). However, the heating is accomplished by using a solution containing sodium sulfide (Na_2S), sodium hydroxide (NaOH), and sodium carbonate (Na_2CO_3) formed from the sulfate during preparation and recovery of the cooking liquor.

An essential factor in the kraft process has been the recovery of the spent liquor from the cooking process. The black liquor removed from the

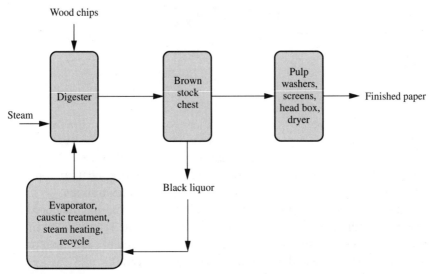

FIGURE 1   The kraft pulping process.

pulp in the pulp washer, or diffuser, contains 95 to 98 percent of the total chemicals charged to the digester. Organic sulfur compounds are present in combination with sodium sulfide. Sodium carbonate is present, as are also small amounts of sodium sulfate, salt, silica, and traces of lime, iron oxide, alumina, and potash. The black liquor is concentrated, burned, and limed to break down any remaining organic compounds. The carbon is burned and the inorganic chemicals melted for reuse or disposal.

PYRIDINE

There are no natural sources of pyridine compounds that are either a single pyridine isomer or just one compound. For instance, coal tar contains a mixture of bases, mostly alkylpyridines, in low concentrations.

Pyridine (boiling point: 115.5°C, density: 0.9819) is manufactured by reacting formaldehyde, acetaldehyde, and ammonia at 350 to 550°C in the presence of a silica-alumina catalyst (SiO_2-Al_2O_3), and the principal products are pyridine (1) and 3-picoline (3-methyl pyridine).

PYROPHOSPHATES

Tetrasodium pyrophosphate (sodium polyphosphate, $Na_4P_2O_7$) is manu-factured by reacting phosphoric acid and soda ash to yield a disodium phosphate solution, which may be dried to give anhydrous disodium phosphate (Na_2HPO_4) or crystallized to give disodium phosphate dihydrate ($Na_2HPO_4 \cdot 2H_2O$) or disodium phosphate heptahydrate ($Na_2HPO_4 \cdot 7H_2O$). These compounds are calcined at a high temperature in an oil-fired or gas-fired rotary kiln to yield tetrasodium pyrophosphate (Fig. 1).

$$2Na_2HPO_4 \rightarrow Na_4P_2O_7 + H_2O$$
$$2Na_2HPO_4 \cdot 2H_2O \rightarrow Na_4P_2O_7 + 5H_2O$$

Tetrasodium pyrophosphate is used as a water softener and as a soap and detergent builder.

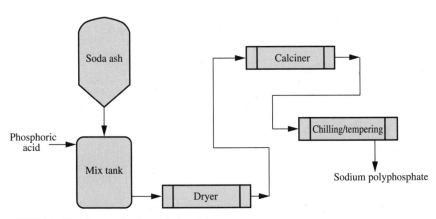

FIGURE 3   Manufacture of sodium polyphosphate.

QUINOLINE

Replacing one carbon atom of naphthalene with a nitrogen atom creates the isomeric quinoline (melting point: −15.6°C, boiling point: 238°C, density: 1.0929) and *iso*-quinoline (melting point: 26.5°C, boiling point: 243°C, density: 1.0986).

Quinoline is isolated from coal-tar distillates in a process in which the tar acids are removed by caustic extraction, and the oil is distilled to produce the methylnaphthalene fraction (230 to 280°C). Washing with dilute sulfuric acid produces sulfate salts, from which the tar bases are liberated by treatment with caustic followed by distillation. The composition of this product is typically 92% quinoline and 5% *iso*-quinoline by weight with smaller amounts of all monomethylquinolines, 2,8-dimethylquinoline, and some homologues of *iso*-quinoline.

Iso-quinoline (>95% pure) can be isolated by treating a crude fraction with hydrochloric acid followed by addition of an alcoholic solution of cupric chloride in a mole ratio of 1:2 $CuCl_2$/*iso*-quinoline.

See Naphthalene.

ISO-QUINOLINE

See **Quinoline.**

RARE GASES

Oxygen (approximately 20% by volume) and nitrogen (approximately 80% by volume) are the primary components of the atmosphere, but air also contains argon, neon, krypton, and xenon (approximately 1% by volume).

Argon, neon, krypton, and xenon are all produced commercially as byproducts from large cryogenic air separation plants. The distillation of liquid air is normally performed in the double-column arrangement (Fig. 1). The rare gases are produced in side columns operated in conjunction with the standard double-column plant.

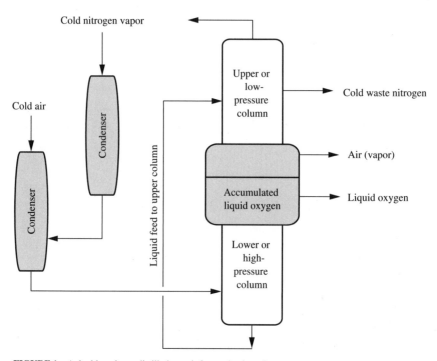

FIGURE 1   A double-column distillation unit for production of rare gases.

Since *argon* boils at a temperature just below oxygen, its concentration level builds up in the upper column at a point above the oxygen product level. The argon-rich vapor is withdrawn from the upper column and is fed to a side argon column. The liquid reflux from the argon column is returned to the upper column at the same point as the vapor withdrawal. The crude argon product is withdrawn from the top of the argon column. The crude argon, which contains oxygen and nitrogen, is processed further to remove oxygen (by the addition of hydrogen and subsequent catalytic combustion and gas drying to recover the water) and nitrogen (by another distillation step) that produces argon having a purity of 99.999%.

Neon boils at a considerably lower temperature than nitrogen and usually collects in the dome of the main condenser as a noncondensable gas. It can be recovered by the addition of a side column.

Krypton and *xenon* have high boiling points relative to oxygen and tend to accumulate in the liquid oxygen sump of the upper column of the main plant.

Helium can be produced from natural gas. A typical plant removes the 2% helium from natural gas of up to 95 percent. The pipeline gas (at 3 to 4.5 MPa) is scrubbed, to remove water and condensable hydrocarbons and is then passed through a gas cleaner, which removes pipeline dust. From the cleaner, the gas goes to absorption towers to remove carbon dioxide (using monoethanolamine) and finally passes through a bauxite dryer. To obtain the helium, the purified gas enters coolers where the gas is chilled to $-156°C$ and then expanded into a separator-rectifier column. The natural gas is liquefied and separated and the crude-helium (75% helium, 25% nitrogen), passes through a heat exchanger counter to the incoming gas.

The crude helium is purified by removing any trace amounts of hydrogen (using a reactor with a small amount of air, where the hydrogen is oxidized to water over a platinum catalyst) and the hydrogen-free gas is further purified utilizing a pressure-swing adsorption (PSA) process that removes all contaminants to a very low level, usually less than 10 ppm. The pressure-swing adsorption process does not remove neon but for most helium uses it is not considered a contaminant.

RDX

*See **Explosives**.*

RED LEAD

Red lead (Pb_3O_4) has a brilliant red-orange color, is quite resistant to light, and finds extensive use as a priming coat for structural steel because it possesses corrosion-inhibiting properties.

Red lead, or *minium*, is manufactured by oxidizing lead to litharge (PbO) in air and further oxidizing the litharge to red lead. In the fumed process, which produces smaller particles, molten lead is atomized by compressed air and then forced through the center of a gas flame, which in turn converts it into litharge as a fume collected in filter bags. The litharge is then oxidized to red lead by roasting in air.

RESERPINE

Reserpine, an indole alkaloid that is obtained from the *Rauwolfia* plant, was the first successful drug to treat high blood pressure.

Reserpine is isolated from its plant producers by using a nonaqueous solvent process, using, for example, boiling methanol extraction of the African root *Rauwolfia vomitoria*. Naturally, these extractions are carried out under countercurrent methods. The methanol extract is concentrated and acidified with 15% acetic acid and then treated with petroleum naphtha to remove impurities. Extraction is made using ethylene dichloride. The solvent is neutralized with dilute sodium carbonate, evaporated to drive off the ethylene dichloride, and further evaporated to crystallize the crude reserpine crystals that are then crystallized.

In common with other indole derivatives, reserpine is susceptible to decomposition by light and oxidation, so it must be stabilized. Modifying the trimethoxyphenyl portion of the molecule gives other antihypertensive drugs with various potency and rapidity of action.

Reserpine is used for its tranquilizing effect on the cardiovascular and central nervous systems and as an adjunct in psychotherapy.

ROTENONE

Rotenone is the toxic principle of several tropical and subtropical plants, the chief of which is derris. It is a complex organic heterocyclic compound and rotenone derivatives are stomach and contact poisons.

Ground derris roots are extracted with chloroform or carbon tetrachloride and the solvent evaporated, leaving a mixture of rotenone and some other less toxic substances, which are not separated.

RUBBER (NATURAL)

Rubber is a natural polymer that is obtained from the rubber tree and has the all *cis*-1 ,4-polyisoprene structure. This structure has been duplicated in the laboratory and is called *synthetic rubber*, made with the use of Ziegler-Natta catalysis.

Natural rubber may contain less than 10% of nonrubber chemicals and has an outstanding heat-buildup resistance.

RUBBER (SYNTHETIC)

Synthetic rubbers are manufactured from a variety of starting materials that have been classified into vulcanizable and nonvulcanizable and also by the chemical composition of the polymer chain.

The most widely used synthetic rubber is styrene-butadiene rubber (SBR) (Fig. 1). Other commonly used elastomers are polybutadiene, polyethylene-propylene, butyl rubber, neoprene, nitrile rubbers, and polyisoprene.

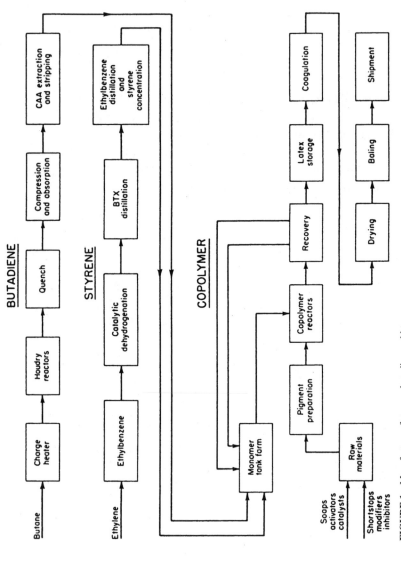

FIGURE 1 Manufacture of styrene–butadiene rubber.

SALICYLIC ACID

The chief derivative of salicylic acid (melting point: 159°C, boiling point: 211°C, density: 1.565), which is used as a drug, is the methyl acetyl ester, known as *aspirin*.

The manufacture of salicylic acid follows carboxylation of sodium phenolate (Fig. 1). The sodium phenolate must be finely divided and exposed to the action of the carbon dioxide under pressure and heat in a heated ball

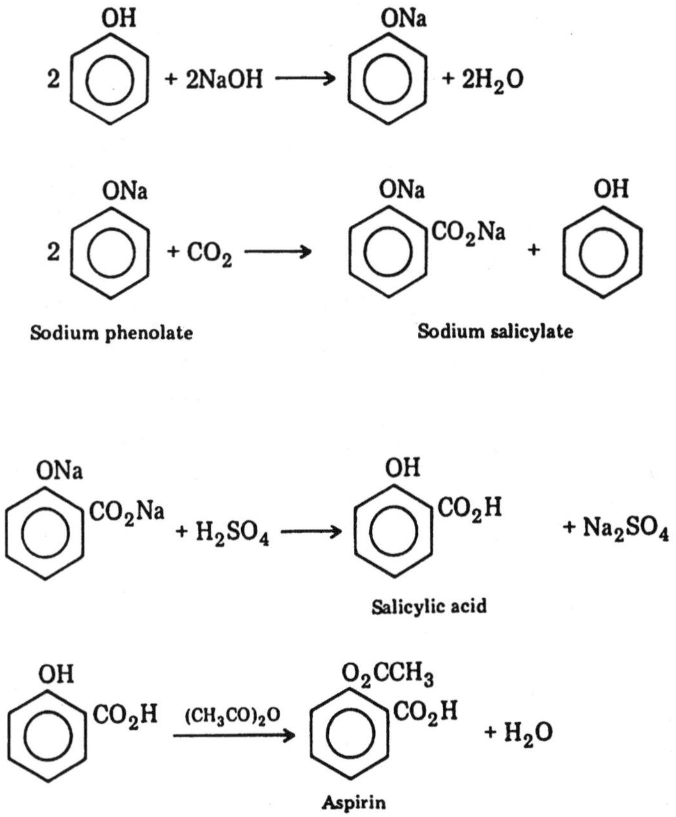

FIGURE 1   Chemistry of salicylic acid and aspirin production.

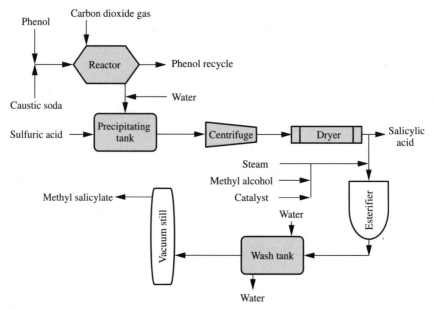

FIGURE 2   Manufacture of salicylic acid and aspirin.

mill reactor (Fig. 2). In the reactor at 130°C and vacuum, the sodium phenolate is reduced to a very dry powder, after which the carbon dioxide is introduced under pressure (700 kPa) and temperature (100°C) to form, first, sodium phenyl carbonate, which isomerizes to sodium salicylate. This can be dissolved out of the mill, and the salicylic acid decolorized by activated carbon and precipitated by addition of sulfuric acid, after which the salicylic acid is purified by sublimation.

To form aspirin, the salicylic acid is refluxed with acetic anhydride in toluene at 88 to 92°C for 20 hours. The reaction mixture is then cooled in aluminum cooling tanks, and the acetylsalicylic acid precipitates as large crystals that are separated by filtration or by centrifugation, washed, and dried.

See Aspirin.

SILICA GEL

See **Sodium Silicate.**

SILVER SULFITE

See Sulfurous Acid.

SOAP

Soaps are the sodium or potassium salts of certain fatty acids obtained from the hydrolysis of triglycerides.

$$Fat + NaOH \rightarrow glycerol + R\text{–}CO_2^{-}Na^{+}$$

Soap comprises the sodium or potassium salts of various fatty acids, but chiefly of oleic, stearic, palmitic, lauric, and myristic acids.

Manufacturing processes are both batch (in which the triglyceride is steam-hydrolyzed to the fatty acid without strong caustic, and then in a separate step it is converted into the sodium salt) or continuous.

The manufacture of soap (Fig. 1) involves continuous splitting (hydrolysis) and, after separation of the glycerin, neutralization of the fatty acids to soap. The procedure is to split, or hydrolyze, the fat, and then, after separation from the glycerol (glycerin) to neutralize the fatty acids with a caustic soda solution:

$$(C_{17}H_{35}COO)_3C_3H_5 + 3H_2O \rightarrow 3C_{17}H35COOH + C_3H_5(OH)_5$$
$$C_{17}H_{35}COOH + NaOH \rightarrow C_{17}H_{35}COONa + H_2O$$

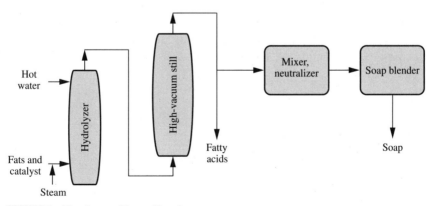

FIGURE 1 Manufacture of fatty acids and soap.

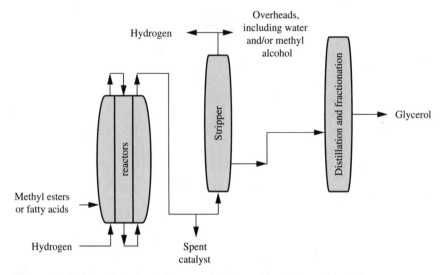

FIGURE 2   Hydrogenolysis of methyl esters to obtain fatty acids and glycerol (glycerin).

In continuous, countercurrent splitting, the fatty oil is deaerated under a vacuum to prevent darkening by oxidation during processing. It is charged at a controlled rate to the bottom of the hydrolyzing tower through a sparge ring (Fig. 2). The oil in the bottom contacting section rises because of its lower density and extracts the small amount of fatty material dissolved in the aqueous glycerol (glycerin) phase. At the same time, deaerated, demineralized water is fed to the top contacting section, where it extracts the glycerol dissolved in the fatty phase. After leaving the contacting sections, the two streams enter the reaction zone where they are brought to reaction temperature by the direct injection of high-pressure steam, and then the final phases of splitting occur. The fatty acids are discharged from the top of the splitter or hydrolyzer to a decanter, where the entrained water is separated or flashed off. The glycerol-water solution is then discharged from the bottom of an automatic interface controller to a settling tank.

SODIUM

Sodium is a silvery-white reactive metal that reacts violently with water and is usually preserved in containers under a nitrogen blanket or under dry, liquid kerosene.

The most important method of preparation of sodium is by the electrolysis of fused sodium chloride.

$$2NaCl \rightarrow 2Na + Cl_2$$

The cell for this electrolysis consists of a closed, rectangular, refractory-lined steel box with a carbon anode and an iron cathode. The anode and cathode are arranged in separate compartments to facilitate the recovery of the sodium and the chlorine. Sodium chloride has a high melting point (804°C), but calcium chloride is added to lower it, and the cell is operated at 600°C. The electrolyte is a eutectic of 33.3% sodium chloride arid 66.7% calcium chloride.

A sodium-calcium mixture collects at the cathode, but the solubility of calcium in sodium decreases with decreasing temperatures so that the heavier calcium crystals, which form as the mixture is cooled, settle back into the bath. The crude sodium is filtered at 105 to 110°C, giving sodium of 99.9% purity that is run molten into a nitrogen-filled tank car and allowed to solidify.

SODIUM BICARBONATE

Sodium bicarbonate (also called bicarbonate of soda or baking soda and mined as the ore nahcolite) can be made by treating soda ash with carbon dioxide and water at about 40°C in a contacting tower.

$$Na_2CO_3 + CO_2 + H_2O \rightarrow 2NaHCO_3$$

The suspension of bicarbonate formed is removed from the bottom of the tower, filtered, and washed on a rotary drum filter. The cake is then centrifuged and dried on a continuous belt conveyor at 70°C. Bicarbonate made in this fashion is about 99.9% pure.

Sodium bicarbonate is widely used in the food industry, in making rubber; in pharmaceuticals; as an antacid; in fire extinguishers, soap and detergents, rug cleaners, animal feeds, and textiles; in leather preparation; in soap, detergent, and paper manufacturing; for flue-gas scrubbing; and for many other diversified small-scale uses.

SODIUM BISULFITE

See Sulfurous Acid.

SODIUM CARBONATE

Sodium carbonate (soda ash) was manufactured by the LeBlanc process (discovered in 1773) for many years in Europe. In this process, salt cake (sodium sulfate) reacts with limestone to give sodium carbonate and a side product, gypsum (calcium sulfate).

$$Na_2SO_4 + CaCO_3 \rightarrow Na_2CO_3 + CaSO_4$$

In 1864, Ernest Solvay, a Belgian chemist, invented his ammonia-soda process (Fig. 1), which has replaced the LeBlanc process.

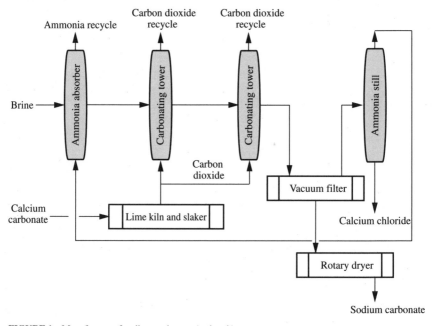

FIGURE 1   Manufacture of sodium carbonate (soda ash).

$$2NH_4OH + 2CO_2 \rightarrow 2NH_4HCO_3$$

$$2NH_4HCO_3 + 2NaCl \rightarrow 2NaHCO_3 + 2NH_4Cl$$

$$2NaHCO_3 \rightarrow Na_2CO_3 + CO_2 + H_2O$$

or

$$CaCO_3 + 2NaCl \rightarrow Na_2CO_3 + CaCl_2$$

In the process (Fig. 1), the brine (salt solution) is mixed with ammonia in a large ammonia absorber. A lime kiln serves as the source of carbon dioxide, which is mixed with the salt and ammonia in carbonation towers to form ammonium bicarbonate and finally sodium bicarbonate and ammonium chloride. Filtration separates the less soluble sodium bicarbonate from the ammonium chloride in solution. The sodium bicarbonate is heated to 175°C in rotary dryers to give light soda ash and the carbon dioxide is recycled. Light soda ash is less dense than the natural material because holes are left in the crystals of sodium bicarbonate as the carbon dioxide is liberated. Dense soda ash, used by the glass industry, is manufactured from light ash by adding water and drying. The ammonium chloride solution goes to an ammonia still where the ammonia is recovered and recycled. The remaining calcium chloride solution is an important by-product of this process, although in large amounts it is difficult to sell and causes a disposal problem.

Natural trona ore is mostly $2Na_2CO_3 \cdot NaHCO_3 \cdot 2H_2O$ (45% Na_2CO_3, 36% $NaHCO_3$, 15% water + impurities). Processing this ore gives soda ash (Fig. 2) and solution mining method is in practice wherever possible.

Glass is the biggest industry using soda ash and consists of bottles and containers, flat glass, and fiberglass. In many other uses, soda ash competes directly with caustic soda as an alkali. The chemical of choice is then dependent on price and availability of the two.

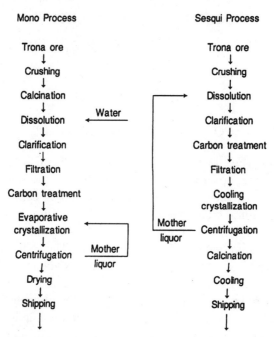

FIGURE 2   Manufacture of sodium carbonate from trona.

SODIUM CHLORATE

Sodium chlorate ($NaClO_3$) is manufactured by the electrolysis of saturated, acidulated brine mixed with sodium dichromate to reduce the corrosive action of the hypochlorous acid present (Fig. 1).

The brine solution is made from soft water or condensate from the evaporator and rock salt purified of calcium and magnesium. The rectangular steel cell is filled with either the brine solution or a recovered salt solution, made from recovered salt-containing chlorate, dissolved in condensate from the evaporator. Electrodes are graphite and steel for small cells, graphite and graphite for larger cells. The temperature of the cell is maintained at 40°C by cooling water.

The electrolysis step produces sodium hydroxide ($NaOH$) at the cathode and chlorine (Cl_2) at the anode, and mixing occurs with the formation of sodium hypochlorite ($NaOCl$) that is oxidized to chlorate.

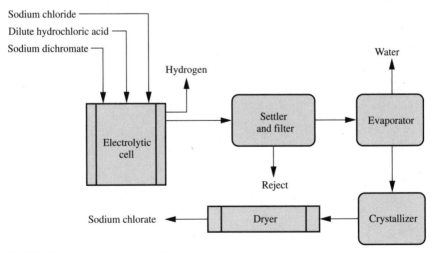

FIGURE 1 Manufacture of sodium chlorate.

$$2NaCl + 2H_2O \rightarrow 2NaOH + H_2 + Cl_2$$

$$Cl_2 + 2NaOH \rightarrow NaOCl + NaCl + H_2O$$

$$3NaOCl \rightarrow NaClO_3 + 2NaCl$$

The cell liquor is pumped to tanks where it is heated with steam to 90°C to destroy any hypochlorite present and the required amount of barium chloride is introduced to precipitate any chromate present.

The graphite mud from the electrodes and the barium chromate settle to the bottom of the tank and the clear liquor is pumped through a filter to the evaporator storage tanks. The liquor in the storage tank is neutralized with soda ash and evaporated, after which the liquor is allowed to settle to remove the sodium chloride. The settled liquid is filtered and cooled and the crystals of sodium chlorate that drop out are separated and dried.

Potassium chloride can be electrolyzed for the direct production of potassium chlorate, but, because sodium chlorate is so much more soluble, the production of the sodium salt is generally preferred. Potassium chlorate may be obtained from the sodium chlorate by a metathesis reaction with potassium chloride.

SODIUM CHLORIDE

Sodium chloride (salt, common salt, rock salt, and grainer salt) is a naturally occurring mineral.

There are three methods of salt production and purification: brine solution, rock salt mining, and the open pan or grainer process.

To produce sodium chloride from *brine*, water is pumped into the salt deposit and the saturated salt solution containing 26% salt, 73.5% water, and 0.5% impurities, is removed. Hydrogen sulfide is removed by aeration and oxidation with chlorine. Calcium (Ca^{2+}), magnesium (Mg^{2+}), and iron (Fe^{3+}) are precipitated as the carbonates using soda ash and are removed in a settling tank. The brine solution can be sold directly or it can be evaporated to give salt of 99.8% purity.

Rock salt is produced from deep *mines* so that the salt is taken directly from the deposit. Salt obtained by this method is 98.5 to 99.4% pure.

In the *open pan* or *grainer salt* method, hot brine solution is held in an open pan approximately 4 to 6 meters wide, 45 to 60 meters long, and 60 cm deep at 96°C. Flat, pure sodium chloride crystals form on the surface and fall to the bottom and are raked to a centrifuge, separated from the brine, and dried. A purity of 99.98% is obtained. A *vacuum pan* system (Fig. 1) is also available.

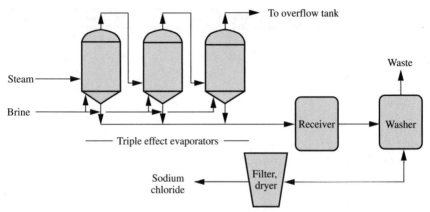

FIGURE 1   Vacuum pan system for producing sodium chloride.

SODIUM CHLORITE

Sodium chlorite ($NaClO_2$) is manufactured from chlorine dioxide, sodium hydroxide, and calcium hydroxide.

$$4NaOH + Ca(OH)_2 + C + 4ClO_2 \rightarrow 4NaClO_2 + CaCO_3 + 3H_2O$$

After filtering oil, the calcium carbonate, the solution of sodium chlorite ($NaClO_2$) is evaporated and drum-dried.

Sodium chlorite is a powerful and stable oxidizing agent and is capable of bleaching much of the coloration in cellulosic materials without weakening the cellulose fibers. It finds uses in the pulp and textile industries, particularly in the final whitening of kraft paper.

Besides being employed as an oxidizer, sodium chlorite is also the source of another chlorine compound, chlorine dioxide.

$$2NaClO_2 + Cl_2 \rightarrow 2NaCl + 2ClO_2$$

Chlorine dioxide has $2^1/_2$ times the bleaching power of chlorine and is important in water purification, for odor control, and for pulp bleaching.

SODIUM DICHROMATE

Sodium dichromate is manufactured from chromite, a chromium iron oxide containing approximately 50% chromic oxide (Cr_2O_3) with iron oxide (FeO), alumina (Al_2O_3), silica (SiO_2), and magnesium oxide (MgO).

In the process (Fig. 1), the ore is ground and mixed with ground limestone and soda ash, and roasted at approximately 1200°C in an oxidizing atmosphere. The sintered mass is crushed and leached with hot water to separate the soluble sodium chromate. The solution is treated with sulfuric acid to convert the sodium chromate to sodium dichromate plus sodium sulfate. Some of the sodium sulfate crystallizes in the anhydrous state from the hot solution during acidification as well as in the evaporators during concentration of the dichromate solution. From the evaporator, the hot, sat-

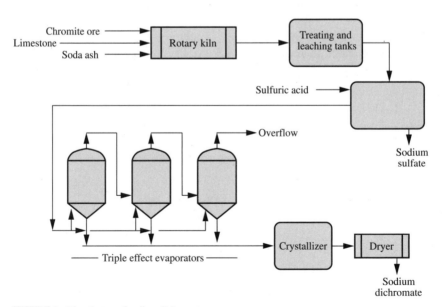

FIGURE 1 Manufacture of sodium dichromate.

urated dichromate solution is fed to the crystallizer, and then to the centrifuge and dryer.

Sodium dichromate is used as the starting material for producing the solutions of chromium salts employed in chrome leather tanning and in chrome mordant dyeing of wool cloth. Pigments, such as yellow lead chromate, are manufactured from sodium dichromate, as are also green chromium oxides for ceramic pigments. Other uses of sodium dichromate include the manufacture of chromium alloys and chromium plating of metals.

SODIUM HYDROXIDE

Sodium hydroxide (caustic soda, caustic) was made for many years by the *lime causticization* method, which involves reaction of slaked lime and soda ash.

$$Na_2CO_3 + Ca(OH)_2 \rightarrow 2NaOH + CaCO_3$$

In 1892, the electrolysis of brine was discovered as a method for making both sodium hydroxide and chlorine, and since the 1960s it has been the only method of manufacture of sodium hydroxide (Fig. 1).

$$2NaCl + 2H_2O \rightarrow 2NaOH + H_2 + Cl_2$$

The brine that is used for the electrolysis must be purified, and calcium, magnesium, and sulfate ions are removed by precipitation reactions.

$$Na_2CO_3 + CaCl_2 \rightarrow CaCO_3 + 2NaCl$$

$$2NaOH + MgCl_2 \rightarrow Mg(OH)_2 + 2NaCl$$

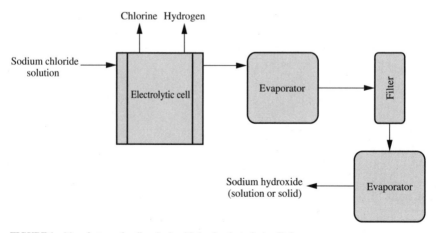

FIGURE 1   Manufacture of sodium hydroxide by the electrolysis of brine.

$$BaCl_2 + Na_2SO_4 \rightarrow BaSO_4 + 2NaCl$$

Two types of cells are employed for the production of sodium hydroxide by electrolysis: the diaphragm cell (Fig. 2) the mercury cell (Fig. 3).

The diaphragm in the diaphragm cell (Fig. 2) prevents the diffusion of sodium hydroxide toward the anode. The anode solution level is maintained higher than in the cathode compartment to retard hack migration. If sodium hydroxide built up near the anode it would react with chlorine to give sodium hypochlorite as a side product.

$$Cl_2 + 2NaOH \rightarrow NaOCl + NaCl + H_2O$$

Each cell is upward of 6 ft square and may contain 100 anodes and cathodes and a sodium hydroxide plant would have several circuits with approximately 90 cells in each circuit.

The mercury cell (Fig. 3) has no diaphragm but is made of two separate compartments. In the electrolyzing chamber the dimensionally stable anodes of ruthenium-titanium cause chloride ion oxidation that is identical to that of a diaphragm cell. The cathode is made of a sodium amalgam flowing across the steel bottom of the cell at a slight angle from the horizontal and promotes the reduction of sodium ions to the metal. The sodium

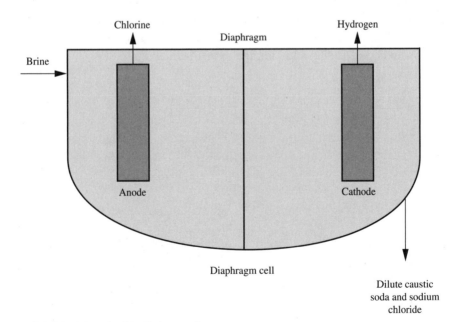

FIGURE 2   Schematic of the diaphragm cell.

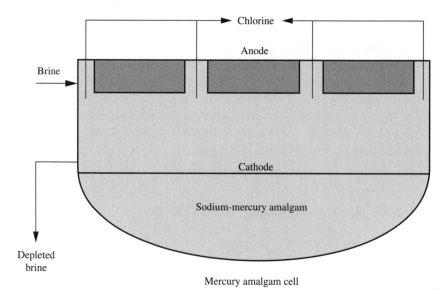

Mercury amalgam cell

FIGURE 3   Schematic of the mercury cell.

amalgam enters a separate denuding chamber where the sodium metal reacts with water. Thus, the overall reaction is identical to that of the diaphragm cell.

Sodium hydroxide has diverse uses and is a reactant in organic and inorganic chemical manufacturing processes. It is also used in the petroleum, pulp and paper, textile, and alumina industries.

SODIUM HYPOCHLORITE

The most common method for manufacturing sodium hypochlorite is by the treatment of sodium hydroxide solution with gaseous chlorine.

$$2NaOH + Cl_2 \rightarrow 2NaOCl + NaCl + H_2O$$

Sodium hypochlorite is employed as a disinfectant and deodorant in dairies, creameries, water supplies, sewage disposal, and households. It is also used as bleach in laundries. As a bleaching agent, it is very useful for cotton, linen, jute, rayon, paper pulp, and oranges.

SODIUM METABISULFITE

See **Sulfurous Acid.**

SODIUM NITRATE

Sodium nitrate (Chile saltpeter, $NaNO_3$) occurs naturally in the highlands of Chile and countercurrent leaching and crystallization produces a good-quality product.

Sodium nitrate is also manufactured from salt (NaCl) or soda ash (Na_2CO_3) and nitric acid.

$$NaCl + HNO_3 \rightarrow NaNO_3 + HCl$$
$$Na_2CO_3 + 2HNO_3 \rightarrow 2NaNO_3 + H_2O + CO_2$$

Sodium nitrate is used in fertilizers, fluxes, fireworks, pickling, and heat-treating mixes and as a tobacco additive.

SODIUM PERCHLORATE

*See **Potassium Perchlorate.***

SODIUM PHOSPHATE

The various sodium phosphates include monosodium phosphate (NaH_2PO_4), disodium phosphate (Na_2HPO_4), and trisodium phosphate ($4Na_3PO_4$ NaOH $48H_2O$).

The first two sodium salts are made from phosphoric acid and soda ash reacted in the proper molecular proportions; the solution is purified if necessary, evaporated, dried, and milled. Trisodium phosphate is also made from phosphoric acid and soda ash, but caustic soda is necessary to substitute the third hydrogen of the phosphoric acid.

To produce sodium tripolyphosphate, a definite temperature control is necessary. When monosodium phosphate and disodium phosphate in correct proportions, or equivalent mixtures of other phosphates, are heated for a substantial time between 300 and 500°C and slowly cooled, the product is practically all in the form of the tripolyphosphate.

$$NaH_2PO_4 + 2Na_2HPO_4 \rightarrow Na_5P_3O_{10} + 2H_2O$$

These salts are employed in water treatment, baking powder (monosodium phosphate), fireproofing, detergents, cleaners, and photography.

SODIUM PYROSULFITE

*See **Sulfurous Acid.***

SODIUM SILICATE

Sodium silicate (silica gel, water glass) is produced when sodium carbonate (soda ash, Na_2CO_3) is heated with sand at 1200 to 1400°C to form various forms of sodium silicate (Fig. 1).

$$Na_2CO_3 + nSiO_2 \rightarrow Na_2O \cdot nSiO_2 + Co$$

Silica gel with a large surface area is used for catalysis and column chromatography. Silica gel is also used as a partial phosphate replacement in soaps and detergents.

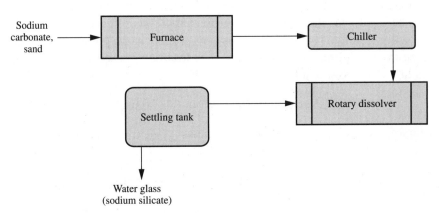

FIGURE 1 Manufacture of sodium silicate.

SODIUM SULFATE

Sodium Sulfate (Na_2SO_4: salt cake; $Na_2SO_4 \cdot 10H_2O$: Glauber's salt) is obtained from a variety of sources.

Manufacture by the Mannheim process involves the reaction of sodium chloride and sulfuric acid at very high temperatures (800 to 900°C).

$$2NaCl + H_2SO_4 \rightarrow Na_2SO_4 + 2HCl$$

However, the majority of sodium sulfate is now obtained directly from natural salt sources. Brines with 7 to 11% sodium sulfate are used and pumped through a salt deposit to lower the solubility of the sodium sulfate so that, upon cooling, the decahydrate (Glauber's salt) will crystallize and can be separated. Heating then forms the anhydrous salt cake.

Sodium sulfate is also obtained as a by-product in the production of viscose rayon. Sulfuric acid and sodium hydroxide are used to degrade the cellulose to rayon in a fiber-spinning bath.

$$2NaOH + H_2SO_4 \rightarrow Na_2SO_4 + 2H_2O$$

Sodium dichromate manufacture also produces sodium sulfate as a by-product.

$$2Na_2CrO_4 + H_2SO_4 + H_2O \rightarrow Na_2Cr_2O_7 + 2H_2O + Na_2SO_4$$

Manufacture by the Hargreaves method also accounts for signifcant sodium sulfate production.

$$4NaCl + 2SO_2 + 2H_2O + O_2 \rightarrow 2Na_2SO_4 + 4HCl$$

Current uses of sodium sulfate include detergents, kraft sulfate pulping, and glass. The percentage of salt cake used in the kraft pulping digestion process has been steadily falling because of a trend away from this method of making paper products. At the same time the amount used in detergents as a phosphate substitute has been increasing.

SODIUM SULFITE

*See **Sulfurous Acid.***

SODIUM TRIPHOSPHATE

Sodium triphosphate (sodium tripolyhosphate) is manufactured by mixing phosphoric acid and sodium carbonate (soda ash) in the calculated amounts to give a 1:2 ratio of monosodium and disodium phosphates and then heating to effect dehydration at 300 to 500°C.

$$2H_3PO_4 + Na_2CO_3 \rightarrow 2NaH_2PO_4 + H_2O + CO_2$$
$$4H_3PO_4 + 4Na_2CO_3 \rightarrow 4Na_2HPO_4 + 4H_2O + 4CO_2$$
$$NaH_2PO_4 + 2Na_2HPO_4 \rightarrow Na_5P_3O_{10} + 2H_2O$$

Sodium triphosphate is used almost solely in one type of product: detergents. Some detergents contain up to 50% by weight sodium triphosphate. It has the unique property of complexing or sequestering dipositive ions such as calcium (Ca^{2+}) and magnesium (Mg^{2+}) that are present in water. Tide® is an example of a phosphate detergent that has been used for at least 5 decades.

Phosphates are prime nutrients for algae and for this reason contribute to greening, eutrophication, and fast aging of lakes.

STEROIDS

The term *steroid* is a general term for a large number of naturally occurring materials found in many plants and animals. Their general structure includes a fused set of three cyclohexanes and one cyclopentane.

Steroid drugs (Fig. 1) include anti-inflammatory agents, sex hormones, and synthetic oral contraceptives. Although the sex hormones are the molecules mainly responsible for differentiating the sexes, it is amazing how similar the male and female hormones are in chemical structure. The only difference between testosterone (male) and progesterone (female) is a hydroxyl (−OH) group versus an acetyl (−CO·OR) group.

Other important steroids are cholesterol and cortisone and the *adrenal cortex hormones*. The adrenal glands secrete more than 50 different steroids, the most important of which are aldosterone and hydrocortisone.

The production of steroids is dependent on: (1) isolation of steroids from natural sources in acceptable yields, (2) conversion into other steroids

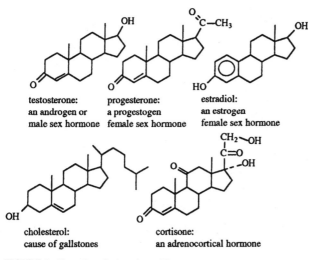

testosterone:
an androgen or
male sex hormone

progesterone:
a progestogen
female sex hormone

estradiol:
an estrogen
female sex hormone

cholesterol:
cause of gallstones

cortisone:
an adrenocortical hormone

FIGURE 1 Formulas of selected steroids.

with the aid of microbial oxidation reactions, and (3) modification with organic synthetic reactions.

The bulk of the world's supply of steroid starting material is derived from two species of plant, the Mexican yam and the soybean. Diosgenin is isolated from the yam in large amounts and treatment with acetic anhydride opens the spiran ring and also acetylates the C-3 hydroxyl (Fig. 2). Oxidation of the newly formed double bond with chromium trioxide makes the desired acetyl group at C-17 of compound and treatment with acetic acid hydrolyzes the ester to a hydroxyl at C-16, which then dehydrates to the double bond to produce 1,6-dehydropregnenolone acetate. Selective catalytic hydrogenation of the new double bond follows to give pregnenolone acetate. The acetate at C-3 is removed by basic hydrolysis to a hydroxyl group, which is then oxidized with aluminum isopropoxide (the Oppenauer reaction) to a keto group. The basic reaction conditions isomerize the double bond so that progesterone, an α,β-unsaturated ketone, is formed. Other routes to progesterone are commercially used, but this is representative.

Large-scale production of cortisone (Fig. 3) from progesterone starts with a microbiologic oxidation with a soil organism, *Rhizopus arrhizus*, to

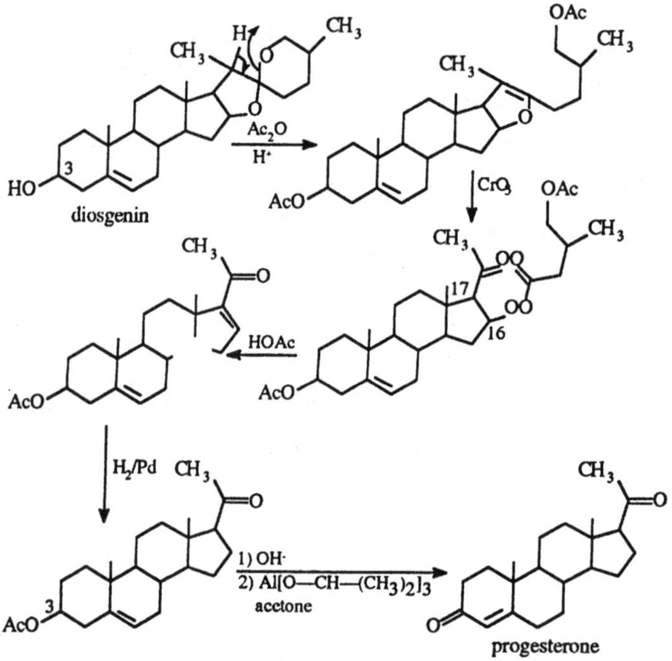

FIGURE 2 Production of progesterone.

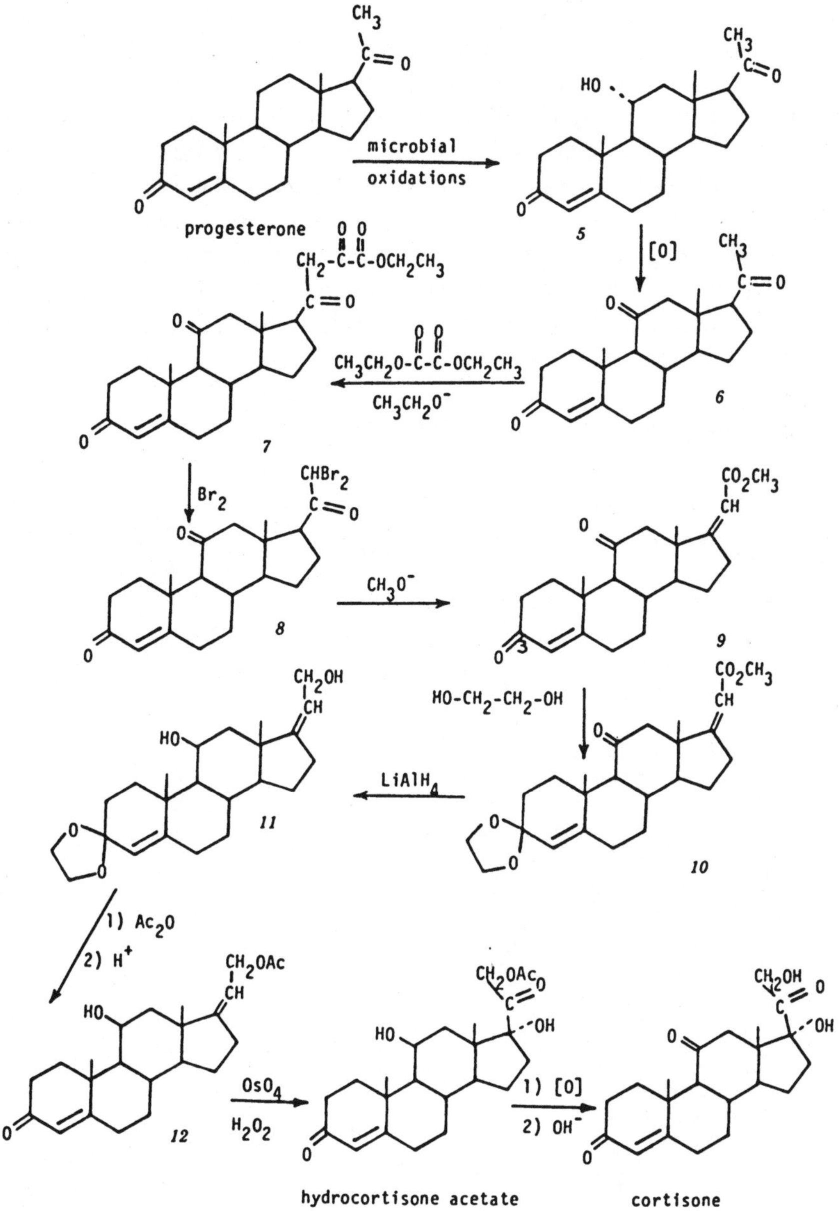

FIGURE 3 Production of cortisone.

convert progesterone into 11 α-hydroxyprogesterone after which oxidation leads to the trione. Condensation with ethyl oxalate activates the appropriate carbon toward selective bromination to form the dibromide. Rearrangement followed by dehydrohalogenation is the next step, and the ketone at C-3 is protected as its ketal. Reaction with lithium aluminum hydride reduces the ester and the C-11 ketone to the alcohol. Acetylation of one of the alcohol groups (the less-hindered primary alcohol) and removal of the protecting group at C-3 then gives the unsaturated acetate, and osmium tetroxide and hydrogen peroxide oxidize the double bond to give hydrocortisone acetate, after which oxidation of the alcohol group and hydrolysis of the acetate gives cortisone.

STREPTOMYCIN

The commercial method for producing this compound involves aerobic submerged fermentation.

The structure of streptomycin indicates its highly hydrophilic nature, and it cannot be extracted by normal solvent procedures. Because of the strong-base characteristics of the two substituted guanidine groups, it may be treated as a cation and removed from the filtered solution by ion-exchange techniques.

STYRENE

Styrene (phenyl ethylene, vinyl benzene; freezing point: –30.6°C, boiling point: 145°C, density: 0.9059, flash point: 31.4°C) is made from ethylbenzene by dehydrogenation at high temperature (630°C) with various metal oxides as catalysts, including zinc, chromium, iron, or magnesium oxides coated on activated carbon, alumina, or bauxite (Fig. 1). Iron oxide on potassium carbonate is also used.

$$C_6H_5CH_2CH_3 \rightarrow C_6H_5CH=CH_2 + H_2$$

Most dehydrogenations do not occur readily even at high temperatures. The driving force for this reaction is the extension in conjugation that results, since the double bond on the side chain is in conjugation with the ring. Conditions must be controlled to avoid polymerization of the styrene and sulfur may be added to prevent polymerization. The crude product is a mixture of styrene, and ethylbenzene that is separated by vacuum distillation, after which the ethylbenzene is recycled. Usually a styrene plant is combined with an ethylbenzene plant when designed.

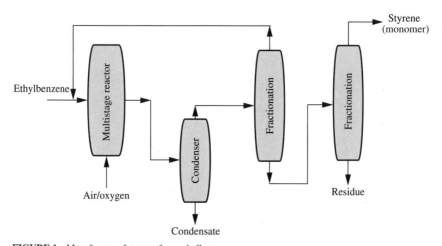

FIGURE 1 Manufacture of styrene from ethylbenzene.

An alternative method for the manufacture of styrene (the oxirane process), uses ethylbenzene that is oxidized to the hydroperoxide and reacts with propylene to give phenylmethylcarbinol (or methyl benzyl alcohol) and propylene oxide. The alcohol is then dehydrated at relatively low temperatures (180 to 400°C) by using an acidic silica gel (SiO_2) or titanium dioxide (TiO_2) catalyst.

$$C_6H_5CH_2CH_3 \rightarrow C_6H_5CH(OOH)CH_3$$

$$C_6H_5CH(OOH)CH_3 + CH_3CH=CH_2 \rightarrow C_6H_5CH(OH)CH_3 + CH_3CH_2CH_2O$$

$$C_6H_5CH(OH)CH_3 \rightarrow C_6H_5CH=CH_2 + H_2O$$

Other methods, such as the direct reaction of benzene and ethylene (Fig. 2) or from pyrolysis gasoline (Fig. 3) are also used to manufacture styrene.

The uses of styrene are dominated by polymer chemistry and involve polystyrene and its copolymers as used in various molded articles such as toys, bottles, and jars and foam for insulation and cushioning.

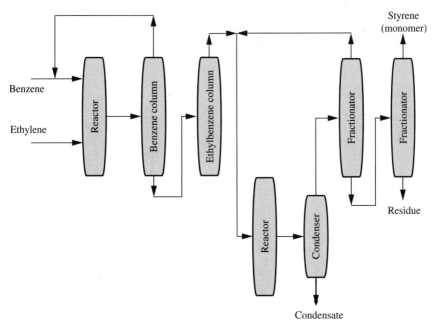

FIGURE 2   Manufacture of styrene from benzene and ethylene.

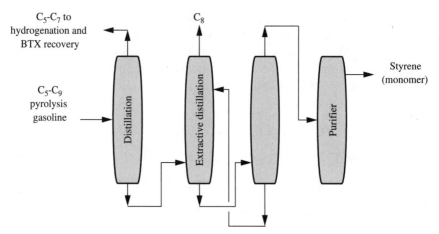

FIGURE 3   Manufacture of styrene from a gasoline fraction.

SULFONAMIDES

The physiologically active sulfonamide (sulfa) drugs involve variations of groups in place of the hydrogen of the sulfonamide moiety.

In the production of a sulfonamide compound, aniline is protected by acetylation to acetanilide to limit the chlorosulfonylation to the *para*-position. Acetylation deactivates the ring toward multielectrophilic attack. Various amines react with sulfonyl chloride to give acetylated sulfonamides. Hydrolysis then removes the acetyl group to give the active drug. Sometimes the drug is administered as its sodium salt, which is soluble in water.

Some common sulfa drugs have changes in the R, with sulfadiazine being probably the best for routine use. It is 8 times as active as sulfanilamide and exhibits fewer toxic reactions than most of the sulfonamides. Most of the common derivatives have an R group that is heterocyclic because of the greater absorption into the body but easier hydrolysis to the active sulfanilamide.

Sulfonamide compounds, although largely replaced by other, newer antibacterial compounds, are still used in treatment of certain infections.

SULFUR

Sulfur occurs naturally in the free state and in ores such as pyrite (FeS_2), sphalerite (ZnS), and chalcopyrite ($CuFeS_2$).

Sulfur is recovered from natural sources such as calcite by the Frasch process (Fig. 1).

Sulfur is also a constituent of petroleum and natural gas (as H_2S). Thus, removing hydrogen sulfide from natural and refinery gases with absorbents such as monoethanolamine and/or diethanolamine also produces elemental sulfur. The hydrogen sulfide is then converted to elemental sulfur by the Claus or modified Claus process (Fig. 2).

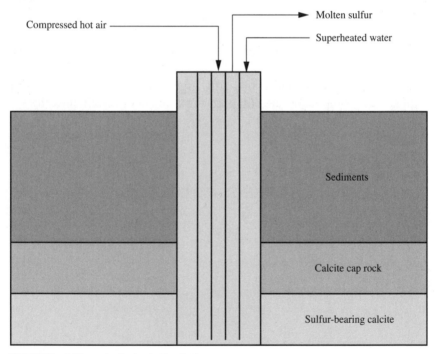

FIGURE 1   Sulfur production by the Frasch process.

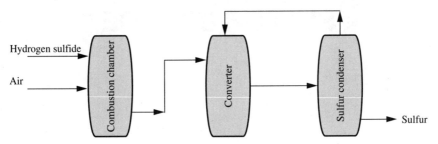

FIGURE 2   Sulfur production by the Claus process.

$$2H_2S + 3O_2 \rightarrow 2H_2O + 2SO_2$$
$$2SO_2 + 2H_2S \rightarrow 3S + H_2O$$

Although there are diverse uses for sulfur, the largest application is in the manufacture of sulfuric acid.

SULFUR DIOXIDE

Sulfur dioxide (boiling point, $-10°C$) is a gas that occurs as a result of the oxidation of sulfur as, for example, during combustion of sulfur-containing fuels.

Sulfur dioxide is manufactured as part of the contact process for making sulfuric acid. Sulfur and oxygen are burned at $1000°C$ ($1830°F$). With very careful control of the amount of air entering the combustion chamber, sulfur dioxide can be produced up to 18% by volume at a temperature of $1200°C$. As the gases from the combustion chamber pass through the heat exchanger, they heat the water for the boilers. The cooled gases, containing from 16 to 18% sulfur dioxide, are pumped into the absorbers through acidproof pumps. The temperature of the vapors coming from the steaming tower depends upon its design, but usually runs about $70°C$. The vapors are cooled and passed through a drying tower in which 98% sulfuric acid is used, although other drying agents may be employed. The sulfur dioxide is liquefied by compression and cooling.

Sulfur dioxide is used for refrigeration and also serves as raw material for the production of sulfuric acid. It is also used as a bleaching agent in the textile and food industries. It is an effective disinfectant and is employed as such for wooden kegs and barrels and brewery apparatus and for the prevention of mold in the drying of fruits. Sulfur dioxide efficiently controls fermentation in the making of wine. It is used in the sulfite process for paper pulp, as a liquid solvent in petroleum refining, and as a raw material in many plants in place of sulfites, bisulfites, or hydrosulfites.

See Sulfuric Acid.

SULFURIC ACID

Sulfuric acid (oil of vitriol, H_2SO_4) is a colorless, oily liquid, dense, highly reactive, and miscible with water in all proportions. Heat is evolved when concentrated sulfuric acid is mixed with water and, as a safety precaution, the acid should be poured into the water rather than water poured into the acid. Anhydrous, 100% sulfuric acid, is a colorless, odorless, heavy, oily liquid (boiling point: 338°C with decomposition to 98.3% sulfuric acid and sulfur trioxide). Oleum is excess sulfur trioxide dissolved in sulfuric acid. For example, 20% oleum is a 20% sulfur trioxide–80% sulfuric acid mix. Sulfuric acid will dissolve most metals and the concentrated acid oxidizes, dehydrates, or sulfonates most organic compounds, sometimes causing charring.

The manufacture of sulfuric acid by the *lead chamber process* involves oxidation of sulfur to sulfur dioxide by oxygen, further oxidation of sulfur dioxide to sulfur trioxide with nitrogen dioxide, and, finally, hydrolysis of sulfur trioxide.

$$S + O_2 \rightarrow SO_2$$

$$2NO + O_2 \rightarrow 2NO_2$$

$$SO_2 + NO_2 \rightarrow SO_3 + NO$$

$$SO_3 + H_2O \rightarrow H_2SO_4$$

Modifications of the process include towers to recover excess nitrogen oxides and to increase the final acid concentration from 65% (*chamber acid*) to 78% (*tower acid*).

The *contact process* has evolved to become the method of choice for sulfuric acid manufacture because of the ability of the process to produce stronger acid.

$$S + O_2 \rightarrow SO_2$$

$$2SO_2 + O_2 \rightarrow 2SO_3$$

$$SO_3 + H_2O \rightarrow H_2SO_4$$

In the process (Fig. 1), sulfur and oxygen are converted to sulfur dioxide at 1000°C and then cooled to 420°C. The sulfur dioxide and oxygen enter the converter, which contains a catalyst such as vanadium pentoxide (V_2O_5). About 60 to 65% of the sulfur dioxide is converted by an exothermic reaction to sulfur trioxide in the first layer with a 2 to 4-second contact time. The gas leaves the converter at 600°C and is cooled to 400°C before it enters the second layer of catalyst. After the third layer, about 95% of the sulfur dioxide is converted into sulfur trioxide. The mixture is then fed to the initial absorption tower, where the sulfur trioxide is hydrated to sulfuric acid after which the gas mixture is reheated to 420°C and enters the fourth layer of catalyst that gives overall a 99.7% conversion of sulfur dioxide to sulfur trioxide. It is cooled and then fed to the final absorption tower and hydrated to sulfuric acid. The final sulfuric acid concentration is 98 to 99% (1 to 2% water). A small amount of this acid is recycled by adding some water and recirculating into the towers to pick up more sulfur trioxide.

Although sulfur is the common starting raw material, other sources of sulfur dioxide can be used, including iron, copper, lead, nickel, and zinc sulfides. Hydrogen sulfide, a by-product of petroleum refining and natural

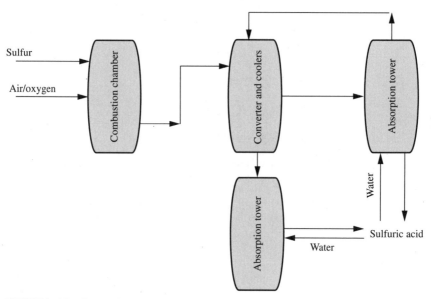

FIGURE 1 Manufacture of sulfuric acid by the contact process.

gas refining, can be burned to sulfur dioxide. Gypsum ($CaSO_4$) can also be used but needs high temperatures to be converted to sulfur dioxide. Other uses for sulfuric acid include the manufacture of fertilizers, chemicals, inorganic pigments, petroleum refining, etching, as a catalyst in alkylation processes, in electroplating baths, for pickling and other operations in iron and steel production, in rayon and film manufacture, in the making of explosives, and in nonferrous metallurgy

SULFUROUS ACID

Sulfurous acid (H_2SO_3) is a colorless liquid, prepared by dissolving sulfur dioxide (SO_2) in water. Reagent grade sulfurous acid contains approximately 6% sulfur dioxide in solution.

Sodium sulfite (Na_2SO_3) and sodium bisulfite (sodium hydrogen sulfite, $NaHSO_3$) are formed by the reaction of sulfurous acid and sodium hydroxide (NaOH) or sodium carbonate (Na_2CO_3) in the proper proportions and concentrations. When dry, on heating sodium sulfite yields sodium sulfate (Na_2SO_4) and sodium sulfide (Na_2S).

Crystalline sulfites are obtained by warming the corresponding bisulfite solutions. Calcium hydrogen sulfite [$Ca(HSO_3)_2$] is used in conjunction with excess sulfurous acid in converting wood to paper pulp. Sodium sulfite and silver nitrate solutions react to yield silver sulfite (Ag_2SO_3), a white precipitate, which upon boiling decomposes, forming silver sulfide, a brown precipitate.

Sulfurous acid forms dimethyl sulfite ($(CH_3O)_2SO$, boiling point: 126°C) with methyl alcohol and diethyl sulfite [$(C_2H_5O)_2SO$, boiling point: 161°C] with ethyl alcohol. As a bleaching agent, sulfurous acid is used for whitening wool, silk, feathers, sponge, straw, wood, and other natural products. In some areas, its use is permitted for bleaching and preserving dried fruits.

SULFUR TRIOXIDE

Sulfur trioxide is a pungent gas that is produced by the oxidation of sulfur dioxide or the complete combustion of sulfur with oxygen. It is also manufactured by distillation of strong oleum ($H_2SO_4 \cdot SO_3$).

Liquid sulfur trioxide is used for sulfonation, especially in the manufacture of detergents. In the past, the difficulty was the instability of the sulfur trioxide. However, under the trade name Sulfans®, stabilized forms of sulfur trioxide are available; several patented inhibitors such as boron compounds, methane sulfonyl chloride, sulfur, tellurium, and phosphorus oxychloride inhibit crystallization or conversion to a polymer.

*See **Sulfuric Acid.***

SUPERPHOSPHATES

The acidification of phosphate rock to produce superphosphate is an important method of making phosphate available for fertilizer purposes.

$$CaH_3(PO_4)_2 + 2H_2SO_4 + 4H_2O \rightarrow CaH_4(PO_4)_2 + 2CaSO_4 \cdot 2H_2O$$

$$CaF_2 + H_2SO_4 + 2H_2O \rightarrow CaSO_4 \cdot 2H_2O + 2HF$$

$$4HF + SiO_2 \rightarrow SiF_4 + 2H_2O$$

$$3SiF_4 + 2H_2O \rightarrow SiO_2 + 2H_2SiF_6$$

The manufacture of superphosphate involves:

1. Preparation of phosphate rock
2. Mixing with acid
3. Curing and drying of the original slurry by completion of the reactions
4. Excavation, milling, and bagging of the finished product.

These steps can be performed in stepwise processes or continuous processes.

See **Phosphoric Acid, Sodium Phosphate.**

SURFACTANTS

Surfactants are chemicals that, when dissolved in water or another solvent, orient themselves at the interface between the liquid and a second solid, liquid, or gaseous phase and modify the properties of the interface.

Surfactants, as a chemical class, have a common molecular similarity insofar as part of the molecule has a long nonpolar (frequently hydrocarbon, hydrophobic) chain that promotes oil solubility and water insolubility and a polar (hydrophilic) part. The hydrophobic portion is a hydrocarbon containing 8 to 18 carbon atoms in a straight or slightly branched chain. In certain cases, a benzene ring may replace some of the carbon atoms in the chain. The hydrophilic functional group may vary widely and may be anionic, e.g., SO_3^{2-}, cationic, e.g., $-N(CH_3)_3+$ or C_5H_5N+; or nonionic, e.g., $-(OCH_2CH_2)-OH$.

In the anionic class, the most used compounds are linear alkylbenzene sulfonates from petroleum and alkyl sulfates from animal and vegetable fats. The straight-chain paraffins or olefins needed are produced from petroleum.

Linear olefins are prepared by dehydrogenation of paraffins, by polymerization of ethylene to *a*-olefins using a triethyl aluminum catalyst (Ziegler-type catalyst), by cracking paraffin wax, or by dehydrohalogenation of alkyl halides.

a-olefins or alkane halides can be used to alkylate benzene through the Friedel-Crafts reaction, employing hydrofluoric acid or aluminum fluoride as a catalyst.

The Ziegler catalytic procedure for converting *a*-olefins to fatty alcohols and the methyl ester hydrogenation process are the important methods for preparing fatty alcohols.

Surfactants can be divided into four general areas: cationic surfactants, anionic surfactants, nonionic surfactants, and amphoteric surfactants. Major anionic surfactants are soaps, linear alcohol sulfates, linear alcohol ethoxysulfates, and linear alkylbenzenesulfonates.

*See **Surfactants (Amphoteric)**, **Surfactants (Anionic)**, **Surfactants (Cationic)**, **Surfactants (Nonionic)**.*

SURFACTANTS (AMPHOTERIC)

Amphoteric surfactants carry both a positive and a negative charge in the organic part of the molecule. They still have a long hydrocarbon chain as the hydrophobic tail and behave as anionic surfactants or cationic surfactants, depending on the pH.

Amphoteric surfactants are used in shampoos and can be used with alkalis for greasy surfaces as well as in acids for rusty surfaces.

See **Surfactants, Surfactants (Anionic), Surfactants (Cationic), Surfactants (Nonionic).**

SURFACTANTS
(ANIONIC)

Anionic surfactants have a molecular structure in which the long hydrophobic alkyl chain is in the anionic part of the molecule.

α-olefin sulfonates are manufactured by the reaction of C_{12}–C_{18} α-olefins with sulfur trioxide followed by reaction with caustic soda. The product is a complex mixture of compounds, and disulfonates are also formed.

Secondary alkanesulfonates are manufactured by the action of sulfur dioxide and air directly on C_{14}–C_{18} n-paraffins (a sulfoxidation reaction), and the sulfonate group can appear in most positions on the chain.

The linear alcohols can be made from other long-chain linear materials, but a process that involves use of a triethylaluminum catalyst allows their formation directly from ethylene and oxygen.

$$nCH_2=CH_2 + O_2\ (+ Et_3Al)\ \rightarrow\ R–CH_2–OH\ (+ Et_3Al)$$

Alcohol ethoxysulfates are made by reaction of 3 to 7 mol of ethylene oxide with a linear C_{12}–C_{14} primary alcohol to give a low-molecular-weight ethoxylate.

Alkyl groups for linear alkylbenzenesulfonate detergents are made through linear α-olefins. n-alkanes can be dehydrogenated to α-olefins, which then can undergo a Friedel-Crafts reaction with benzene as described above for the nonlinear olefins. Sulfonation and basification gives the linear alkylbenzenesulfonate detergent.

Alternatively, linear α-olefins can be made from ethylene by using Ziegler catalysts to give the ethylene oligomer with a double-bonded end group.

$$6CH_2=CH_2\ \rightarrow\ C_{10}H_{21}CH=CH_2$$

Linear alkylbenzenesulfonate detergents made from the chlorination route have lower amounts of 2-phenyl product. Use of the α-olefins gives greater 2-phenyl content, which in turn changes the surfactant action somewhat.

*See **Surfactants, Surfactants (Amphoteric), Surfactants (Cationic), Surfactants (Nonionic)**.*

SURFACTANTS
(CATIONIC)

Cationic surfactants are generally nitrogen compounds and many are quaternary nitrogen compounds, such as tallow fatty acid trimethylammonium chloride. In the more general structure $R^1R^2R^3R^4N^+X^-$, R^1 is a long alkyl chain, the other R moieties may be alkyl or hydrogen, and X^- is halogen or sulfate ion.

The long hydrocarbon chain is derived from naturally occurring fats or triglycerides, that is, triesters of glycerol having long chain acids with an even number of carbons, being of animal or vegetable origin. A common source for cationic surfactants is inedible tallow from meat packing plants. If the fatty acid is required, the ester is hydrolyzed at high temperature and pressure, or with a catalyst such as zinc oxide or sulfuric and sulfonic acid mixtures. The fatty acid is then converted into the quaternary nitrogen salt.

Cationic surfactants have applications such as inhibiting the growth of bacteria, inhibiting corrosion, separating phosphate ore from silica and potassium chloride from sodium chloride (flotation agents), and they serve well as fabric softeners, antistatic agents, and hair conditioners.

See Surfactants, Surfactants (Amphoteric), Surfactants (Anionic), Surfactants (Nonionic).

SURFACTANTS
(NONIONIC)

Nonionic surfactants have a molecular arrangement in which there is a nonpolar hydrophobic portion and a more polar, but not ionic, hydrophilic part capable of hydrogen bonding with water.

The major nonionic surfactants have been the reaction products of ethylene oxide and nonylphenol. Dehydrogenation of n-alkanes from petroleum (C_9H_{20}) is the source of the linear nonene.

They are now being replaced by the polyoxyethylene derivative of straight-chain primary or secondary alcohols with C_{10}–C_{18}. These linear alcohol ethoxylate nonionic surfactants are more biodegradable than nonylphenol derivatives and have better detergent properties than linear alkylbenzenesulfonate.

$$C_{14}H_{29}\text{–OH} + n\,CH_2CH_2\text{–O} \rightarrow C_{14}H_{29}\text{–O}(CH_2CH_2O)_n H$$

*See **Surfactants, Surfactants (Amphoteric), Surfactants (Anionic), Surfactants (Cationic).***

SYNTHESIS GAS

Synthesis gas (syngas) is a mixture of carbon monoxide and hydrogen that is produced from the reaction of carbon (usually coal or coke or similar carbonaceous material) with steam.

$$C + H_2O \rightarrow CO + H_2$$

$$CO + H_2O \rightarrow CO_2 + H_2$$

$$C + CO_2 \rightarrow 2CO$$

There are three reactor types for gasification processes: (1) a gasifier reactor, (2) a devolatilizer, and (3) a hydrogasifier with the choice of a particular design, e.g., whether or not two stages should be involved, depending on the ultimate product gas desired. Reactors may also be designed to operate over a range of pressure from atmospheric to high pressure and gasification processes can also be segregated according to the bed types: (1) fixed bed, (2) moving bed, (3) fluidized bed, and (4) entrained bed.

Purification of synthesis gas is an important aspect of the process and involves the removal of carbon oxides to prevent poisoning of the catalyst. An absorption process (ethanolamine or hot carbonate) is used to remove the bulk of the carbon dioxide, followed by methanation of the residual carbon oxides in the methanator.

In the production of paraffins, the mixture of carbon monoxide and hydrogen is enriched with hydrogen from the water-gas catalytic (Bosch) process, i.e., shift reaction (Fig. 1), and passed over a cobalt-thoria catalyst to form straight chain (linear) paraffins, olefins, and alcohols (Fischer-Tropsch synthesis):

$$nCO + (2n + 1)H_2 \ (+ \text{ cobalt catalyst}) \rightarrow C_nH_{2n+2} + nH_2O$$

$$2nCO + (n + 1)H_2 \ (+ \text{ iron catalyst}) \rightarrow C_nH_{2n+2} + nCO_2$$

$$nCO + 2nH_2 \ (+ \text{ cobalt catalyst}) \rightarrow C_nH_{2n} + nH_2O$$

$$2nCO + nH_2 \ (+ \text{ iron catalyst}) \rightarrow C_nH_{2n} + nCO_2$$

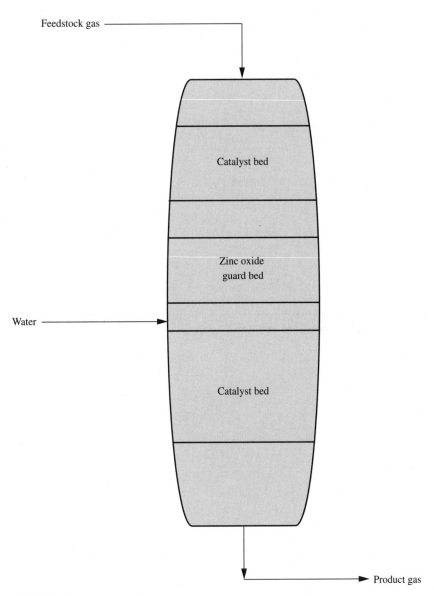

FIGURE 1 Schematic of a shift converter.

$$nCO + 2nH_2 \text{ (+ cobalt catalyst)} \rightarrow C_nH_{2n+1}OH + (n-1)H_2O$$

$$(2n-1)CO + (n+1)H_2 \text{ (+ iron catalyst)} \rightarrow C_nH_{2n+1}OH + (n-1)CO_2$$

Synthesis gas is widely used as a starting material for a variety of chemicals (Fig. 2).

Starting material	Reaction type	Product
Synthesis gas	Oxo reaction	Oxo products
(carbon monoxide	Shift reaction	Hydrogen
+ hydrogen)	Shift reaction	Methyl alcohol
	Shift reaction	Ammonia
	Shift reaction and methanation	Substitute natural gas
	Organic synthesis	Hydroquinone
	Homologation	Ethyl alcohol
	Carbonylation	Acetic acid
	Fischer-Tropsch	Ethylene
	Fischer-Tropsch	Paraffins
	Glycol synthesis	Ethylene glycol

FIGURE 2   Chemicals from synthesis gas.

TALC

The mineral talc is a magnesium silicate [$Mg_3Si_4O_{10}(OH)_2$, density: 2.5 to 2.8] that occurs as foliated to fibrous masses with a waxlike or pearly color, white to gray or green translucent to opaque. It has a distinctly greasy feel.

Talc is found chiefly in the metamorphic rocks, often those of a more basic type because of the alteration of the minerals mentioned above.

A coarse grayish-green talc rock has been called soapstone or steatite and was formerly much used for stoves, sinks, and electrical switchboards. Talc is used as a cosmetic, for lubricants, and as a filler in paper manufacturing. Most tailor's *chalk* consists of talc.

TALL OIL

Tall oil is the generic name for the oil obtained upon acidification of the black liquor residue from kraft pulp digesters. Kraft processing dissolves the fats, fatty acids, rosin, and rosin acids contained in pinewoods in the form of sodium salts and when the black liquor is concentrated to make it possible to recover some of its chemical and heating value, the soaps become insoluble and can be skimmed off. The brown, frothy curd thus obtained is then made acidic with sulfuric acid, converting the constituents to a dark-brown fluid (tall oil).

Tall oil is used as a source of turpentine. Tall oil fatty acids are mostly normal C_{18} acids, 75% mono- and diunsaturated, with lesser amounts of saturated and triunsaturated constituents. Tall oil is also used for waterproofing agents, dimer acids, polyamide resins for printer's ink, adhesives, detergents, and agricultural emulsifiers.

TEREPHTHALIC ACID

Terephthalic acid (boiling point: 300°C) and dimethyl terephthalate (melting point: 141°C) are derived from *p*-xylene by oxidation of *p*-xylene in acetic acid as a solvent in the presence of a variety of catalysts such as cobalt and manganese salts of heavy metal bromides as catalysts at 200°C and 400 psi (Fig. 1).

$$CH_3C_6H_4CH_3 + [O] \rightarrow HOOCC_6H_4COOH$$
$$CH_3C_6H_4CH_3 + [O] \rightarrow CH_3OOCC_6H_4COOCH_3$$

The crude terephthalic acid is cooled and crystallized followed by evaporation of the acetic acid and xylene. The terephthalic acid is washed with hot water to remove traces of the catalyst and acetic acid. If *p*-formylbenzoic acid is present as an impurity from incomplete oxidation, it can be

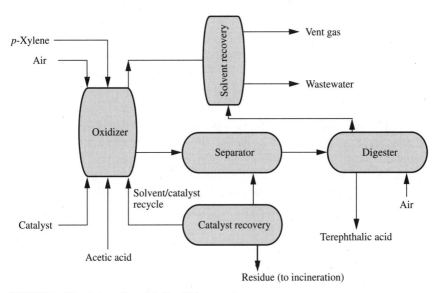

FIGURE 1   Manufacture of terephthalic acid from *p*-xylene.

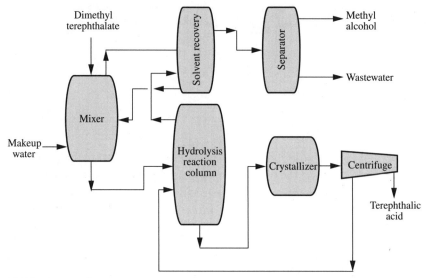

FIGURE 2   Manufacture of terephthalic acid from dimethyl terephthalate.

removed by hydrogenation to *p*-methylbenzoic acid and recrystallization of the pure terephthalic acid, melting point 300°C.

The dimethyl terephthalate can be converted to terephthalic acid by hydrolysis (Fig. 2).

Dimethyl terephthalate is manufactured from terephthalic acid or directly from *p*-xylene. Esterification of terephthalic acid with methanol occurs with sulfuric acid as the acid catalyst. Direct oxidation of *p*-xylene with methanol present also produced dimethyl terephthalate; copper salts and manganese salt are catalysts for this reaction. The dimethyl terephthalate (boiling point 288°C, melting point 141°C) must be carefully purified via a five-column distillation system.

Terephthalic acid and dimethyl terephthalate are used to produce polyester fibers, polyester resins, and polyester film. Terephthalic acid or dimethyl terephthalate is usually reacted with ethylene glycol to give poly(ethylene terephthalate) but sometimes it is combined with 1,4-butanediol to yield poly (butylene terephthalate). Polyester fibers are used in the textile industry. Films find applications as magnetic tapes, electrical insulation, photographic film, packaging, and polyester bottles.

TETRACHLOROETHYLENE

*See **Perchloroethylene.***

TETRACYCLINES

Tetracycline compounds are efficient antibacterial agents and have the broadest effects of any antibacterial discovered.

The tetracyclines are manufactured by fermentation procedures or by chemical modifications of the natural product. Controlled catalytic hydrogenolysis of chlortetracycline, a natural product, selectively removes the 7-chloro atom and produces tetracycline, the most important member of the group.

The hydrochloride salts are used most commonly for oral administration and are usually encapsulated because of their bitter taste.

TETRAHYDROFURAN

Tetrahydrofuran (freezing point: −108°C, boiling point: 67°C, density: 0.8892) can be manufactured from butane by using circulating solids technology in which butane is oxidized to maleic acid and thence to tetrahydrofuran (Fig. 1).

*See **Liquefied Petroleum Gas.***

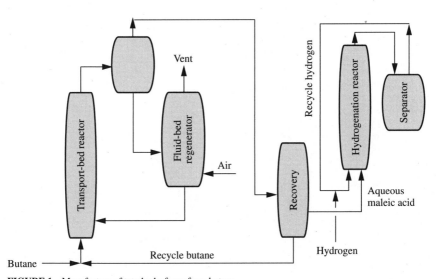

FIGURE 1   Manufacture of tetrahydrofuran from butane.

TETRAZINE

*See **Explosives**.*

TETRYL

Tetryl [2,4,6-trinitrophenylmethylnitramine, $C_6H_2(NO_2)_3NCH_3NO_2$] is manufactured by the action of mixed sulfuric and nitric acid on dimethylaniline in a multiple-stage nitration.

It may also be manufactured by alkylating 2,4-dinitrochlorobenzene with methylamine followed by nitration.

Tetryl is a high explosive with intermediate sensitivity and is used as a base charge in blasting caps, as the booster explosive in high-explosive shells, and as an ingredient of binary explosives.

TITANIUM DIOXIDE

Titanium dioxide (TiO_2, density: 4.26) occurs in two crystalline forms, *anatase* and the more stable *rutile*. Anatase can be converted to rutile by heating to 700 to 950°C. It is variously colored, depending upon source, decomposes at about 1640°C before melting, and is insoluble in water but soluble in sulfuric acid or alkalis.

The two methods for producing titanium dioxide are the *sulfate process* and the *chloride process*. The sulfate process is a batch process introduced by European makers in the early 1930s, and the chloride process, a continuous process, was introduced in the late 1950s. The sulfate process can handle both rutile and anatase, but the chloride process is limited to rutile.

The *sulfate process* (Fig. 1) involves the reaction of ilmenite (an ore containing 45 to 60% by weight titanium dioxide) and treating it with sul-

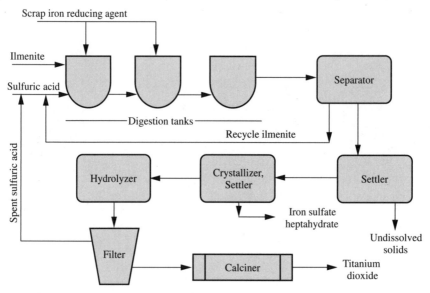

FIGURE 1 Titanium dioxide manufacture by the sulfate process.

furic acid for digestion and filtration. Hydrolysis of the sulfate and heating gives pure titanium dioxide.

$$FeO \cdot TiO_2 + 2H_2SO_4 \rightarrow FeSO_4 + TiO \cdot SO_4 + 2H_2O$$

$$TiO \cdot SO_4 + 2H_2O \rightarrow TiO_2 \cdot H_2O + H_2SO_4$$

$$TiO_2 \cdot H_2O \rightarrow TiO_2 + H_2O$$

The iron sulfate crystallizes out from the titanium persulfate solution and can be recycled to make more sulfuric acid. The sulfate process uses the ore ilmenite as a raw material, while the chloride process requires rutile. Ilmenite can be converted to synthetic rutile.

The sulfate process has traditionally used batch ore digestion, in which concentrated sulfuric acid is reacted with ilmenite. This reaction is very violent and causes the entrainment of sulfur oxides (SO_x) and sulfuric acid in large amounts of water vapor. In an effort to reduce the particulate emissions, scrubbers have been installed at most plants, but these, in turn, have necessitated the treatment of large quantities of scrubbing liquid before discharge. Other waste-disposal problem products are spent sulfuric acid and copperas ($FeSO_4 \cdot 7H_2O$).

The sulfate process has, in some cases, been supplanted by the chloride process because of by-product character and disposal. However, a continuous process that uses relatively dilute sulfuric acid (25 to 60%) to temper the violent, original reaction and to reduce the amount of water-vapor-entrained particulates is available. As the process uses more dilute acid than the older batch process, more of the spent acid can be recycled.

$$TiO_2 \text{ (ore)} + H_2SO_4 \rightarrow TiO \cdot SO_4 + FeSO_4 \cdot H_2O$$

$$TiO \cdot SO_4 + H_2O \rightarrow TiO_2 \cdot xH_2O$$

$$TiO_2 \cdot xH_2O \rightarrow TiO_2 + xH_2O$$

The hydrolysis reaction is dependent upon several factors: quantity and quality of the seeds added to the colloidal suspension of titanium dioxide, concentration, rate of heating, and pH. Introduction of seeds prior to hydrolysis ensures production of the desired form. Using anatase seeds, 6 hours of boiling is needed. and with rutile seeds, the time can be shortened to 3 hours.

The *chloride process* (Fig. 2) involves the reaction of rutile (an ore containing approximately 95% by weight titanium dioxide, TiO_2) with chlorine to give titanium tetrachloride ($TiCl_4$), a liquid that can be purified by

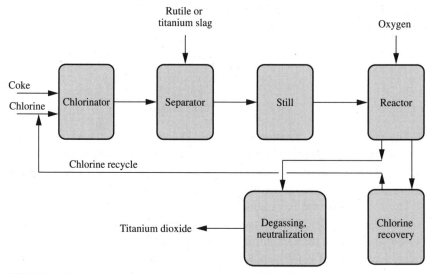

FIGURE 2   Titanium dioxide manufacture by the chloride process.

distillation, boiling point 136°C. The titanium tetrachloride is then oxidized to pure titanium dioxide and the chlorine is regenerated.

$$3TiO_2 + 4C + 6Cl_2 \rightarrow 3TiCl_4 + 2CO + 2CO_2$$

$$TiCl_4 + O_2 \rightarrow TiO_2 + 2Cl_2$$

The chloride process utilizes the treatment of rutile (natural or synthetic) with chlorine gas and coke to produce titanium tetrachloride ($TiCl_4$). The titanium tetrachloride is distilled to remove impurities and then reacted with oxygen or air in a flame at about 1500°C to produce chlorine and very fine particle titanium dioxide. The chlorine is recycled (Fig. 1).

Titanium dioxide is the principal white pigment of commerce. The compound has an exceptionally high refractive index, great inertness, and a negligible color, all qualities that make it close to an ideal white pigment. The major uses of titanium dioxide pigments are: paint, paper, plastics, floor coverings, printing inks, and various applications including rubber, ceramics, roofing granules, and textiles. Almost all of the titanium dioxide used in paints is the rutile form.

TOLUENE

Toluene ($C_6H_5CH_3$, boiling point: 110.8°C, density: 0.8548, flash point: 4.4°C, ignition temperature: 552°C) is a colorless, flammable liquid with a benzenelike odor that is essentially insoluble in water but is fully miscible with alcohol, ether, chloroform, and many other organic liquids. Toluene dissolves iodine, sulfur, oils, fats, resins, and phosgene. When ignited, toluene burns with a smoky flame. Unlike benzene, toluene cannot be easily purified by crystallization.

Toluene is generally produced along with benzene, xylenes, and C_9 aromatics by the catalytic reforming of C_6–C_9 naphthas. The resulting crude reformate is extracted, most frequently with sulfolane (Fig. 1) or tetraethylene glycol and a cosolvent, to yield a mixture of benzene, toluene, xylenes, and C_9–aromatics, which are then separated by fractionation.

The principal source of toluene is catalytic reforming of refinery streams. This source accounts for about 79% of the total toluene produced. An additional 16% is separated from pyrolysis gasoline produced in steam

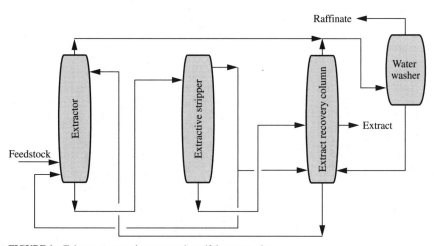

FIGURE 1 Toluene or aromatics recovery by sulfolane extraction.

crackers during the manufacture of ethylene and propylene. Other sources are an additional 1% recovered as a by-product of styrene manufacture and 4% entering the market via separation from coal tars. The reactions taking place in catalytic reforming to yield aromatics are dehydrogenation or aromatization of cyclohexanes, dehydroisomerization of substituted cyclopentanes, and the cyclodehydrogenation of paraffins.

One toluene production process commences with mixed hydrocarbon stocks and can be used for making both toluene and benzene, separately or simultaneously. The process is a combination of extraction and distillation. An aqueous dimethyl sulfoxide (DMSO) solution is passed countercurrently against the mixed hydrocarbon feed. A mixture of aromatic and paraffinic hydrocarbons serves as reflux.

Catalytic reforming is the major source of benzene and xylenes as well as of toluene. There are three basic types of processes: semiregenerative, cyclic, and continuous.

In the semiregenerative process (Fig. 2), feedstocks and operating conditions are controlled so that the unit can be maintained on stream from 6 months to 2 years before shutdown and catalyst regeneration. In cyclic process (Fig. 3), a swing reactor is employed so that one reactor can be regenerated while the other three are in operation. Regeneration, which may be as frequent as every 24 hours, permits continuous operation at high severity. In the continuous process (Fig. 4), the catalyst is continuously withdrawn, regenerated, and fed back to the system.

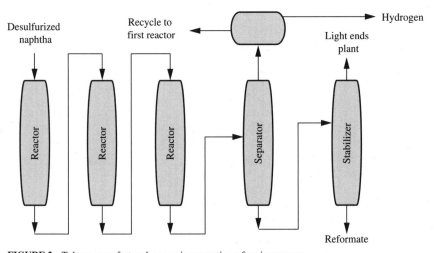

FIGURE 2   Toluene manufacture by a semiregenerative reforming process.

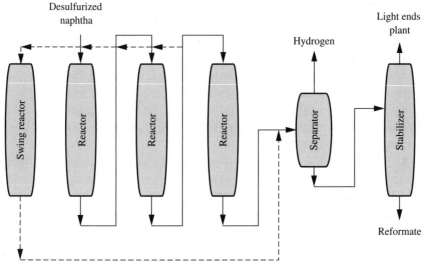

FIGURE 3 Toluene manufacture by a cyclic reforming process.

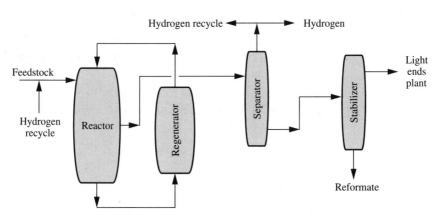

FIGURE 4 Toluene manufacture by a continuous reforming process.

The dealkylation of toluene is a prime source of benzene, accounting for about one-half of toluene consumption. The production of diisocyanates from toluene is increasing. As a component of fuels, the use of toluene is lessening. Toluene takes part in several industrially important syntheses. The hydrogenation of toluene yields methyl cyclohexane ($C_6H_{11}CH_3$), a solvent for fats, oils, rubbers, and waves. Trinitrotoluene [TNT, $CH_3C_6H_2(NO_2)_3$] is a major component of several explosives. When reacted with sulfuric acid, toluene yields o- and p-toluene sulfonic acids

($CH_3C_6H_4SO_3H$). Saccharin is a derivative of the ortho acid; chloramine T (an antiseptic) is a derivative of the para acid.

Like benzene, which is made from toluene by hydrodealkylation, toluene also provides a source for a variety of chemicals (Fig. 5).

Toluene also provides an alternative source for the manufacture of the xylene isomers, especially *p*-xylene. The last two products provide routes respectively to terephthalic acid and *p*-xylene without the need for an isomer separation, a very appealing use for toluene that is often in excess supply, unlike the xylene isomers.

Another process is the conversion of toluene into caprolactam that provides an alternative basic building block for this chemical other than benzene. Toluene is oxidized to benzoic acid, and hydrogenation to cyclohexanecarboxylic acid is followed by treatment with nitrosylsulfuric acid to form cyclohexanone oxime followed by rearrangement to caprolactam.

Two other derivatives of toluene are the explosive trinitrotoluene (TNT) and the polyurethane monomer toluene diisocyanate (TDI).

The production of trinitrotoluene requires complete nitration of toluene that can be achieved by use of nitric acid.

Toluene diisocyanate is derived from a mixture of dinitrotoluenes (usually 65 to 85% *o,p*-dinitrotoluene and 35 to 15% *o,o*-dinitrotoluene) followed by reduction to the diamine and reaction with phosgene to the diisocyanate.

Toluene diisocyanate is made into flexible foam polyurethanes for cushioning in furniture, automobiles, carpets, bedding, polyurethane coatings, rigid foams, and elastomers.

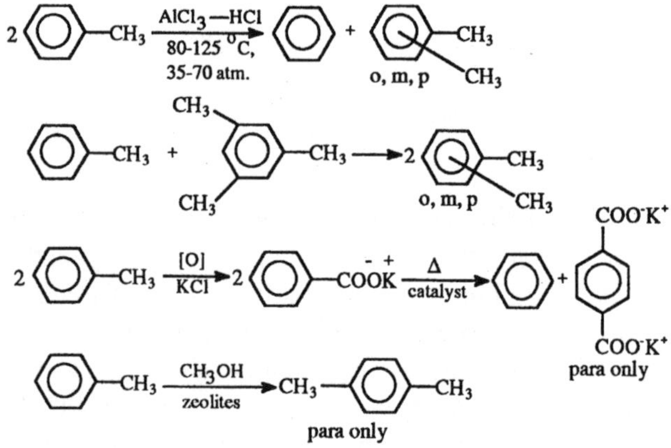

FIGURE 5   Conversion of toluene to other aromatics.

Chlorotoluene ($CH_3C_6H_4Cl$), a widely used solvent for synthetic resins and rubber, is a derivative of toluene. Toluene also is used in the manufacture of benzoic acid ($C_6H_5CO_2H$), the latter an important ingredient for phenol (C_6H_5OH) production.

In other industrially important processes, toluene is a source of benzyl chloride ($C_6H_5CH_2Cl$), benzal chloride ($C_6H_5CHCl_2$), benzotrichloride ($C_6H_5CCl_3$), benzyl alcohol ($C_6H_5CH_2OH$), benzaldehyde (C_6H_5CHO), and sodium benzoate (C_6H_5COONa).

*See **Benzene**.*

TOLUENE DIISOCYANATE

Toluene diisocyanate (TDI) is made from the reaction of 2,4-toluenedi-amine and phosgene. The diamine is made by reduction of dinitrotoluene, which in turn is manufactured by nitration of toluene.

Toluene diisocyanate is used for the production of flexible polyurethane foams for furniture, transportation uses, carpet underlay, and bedding; for coatings; in rigid foams; and elastomers.

1,1,1-TRICHLOROETHANE

1,1,1-trichloroethane (melting point: −30.4°C, boiling point: 74.1°C, density: 1.3390) is made primarily from vinyl chloride by a hydrochlorination-chlorination process.

$$CH_2=CHCl + HCl \rightarrow CH_3CHCl_2$$
$$CH_3CHCl_2 + Cl_2 \rightarrow CH_3CCl_3 + HCl$$

It is also be made from vinylidene chloride by hydrochlorination or from ethane by chlorination.

Uses of 1,1,1-trichloroethane are in vapor degreasing, cold cleaning, aerosols, adhesives, intermediates, and coatings.

TRICHLOROETHYLENE

See **Perchloroethylene.**

TRIETHYLENE GLYCOL

Triethylene glycol is produced, with diethylene glycol, as a by-product in the manufacture of ethylene glycol from hydrolysis of ethylene oxide.

$$6CH_2CH_2O + 3H_2O \rightarrow HOCH_2CH_2OH + HOCH_2CH_2OCH_2CH_2OH$$
$$+ HOCH_2CH_2OCH_2CH_2OCH_2CH_2OH$$

It is separated from the ethylene glycol and diethylene glycol by vacuum distillation.

See **Ethylene Glycol** and **Diethylene Glycol.**

TRINITROTOLUENE

Symmetrical trinitrotoluene (1,3,5-trinitrotoluene, *sym*-trinitrotoluene, TNT) is manufactured by multiple-stage nitration of toluene with a mixture of nitric acid and sulfuric acid.

Three-stage nitration to mono-, di-, and trinitrotoluene was formerly used, but continuous-flow stirred-tank reactors and tubular units using the countercurrent flow of strong acids and toluene permit better yields and reaction control.

TURPENTINE

Turpentine is a mixture of $C_{10}H_{16}$ volatile terpene hydrocarbons (predominantly α-pinene and β-pinene) made of isoprene units).

Turpentine is produced from various species of pines and balsamiferous woods, and several different methods are applied to obtain the oils leading to different types of turpentine, such as (1) dry-distilled wood turpentine from dry distillation of chopped woods and roots of pine trees, (2) steam-distilled wood turpentine that is steam-distilled from pine wood or from solvent extracts of the wood, and (3) sulfate turpentine, which is a by-product of the production of cellulose sulfate.

Pine oil is a mixture of terpine-derived alcohols. It can be extracted from pine but is also synthetically made from turpentine, especially the α-pinene fraction, by reaction with aqueous acid. It is used in many household cleaners as a bactericide, odorant, and solvent.

Rosin, a brittle solid, melting point 80°C, is obtained from the gum of trees and tree stumps as a residue after steam distillation of the turpentine (Fig. 1). It is made up of 90% resin acids and 10% neutral matter. Resin

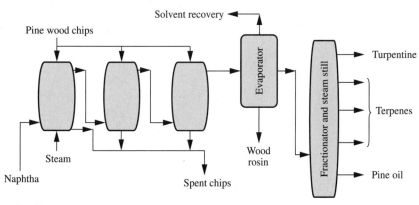

FIGURE 1 Production of turpentine.

acids are tricyclic monocarboxylic acids of formula $C_{20}H_{30}O_2$. The common isomer is 1-abietic acid. About 38% of rosin is used as paper size (its sodium salt), in synthetic rubber as an emulsifier in polymerization (13%), and in adhesives (12%), coatings (8%), and inks (8%).

In addition to turpentine, rosin, and pine oil that can be obtained from pines, directly or indirectly by distillation or extraction, the kraft pulp process now furnishes many related side products. Sulfate turpentine can be obtained from the black kraft liquor. Tall oil rosin and tall oil fatty acids can also be isolated from this liquor. Tall oil rosin is similar to pine rosin and is used in paper sizing, printing inks, adhesives, rubber emulsifiers, and coatings. Tall oil fatty acids are C_{16} and C_{18} long-chain carboxylic acids used in coatings, inks, soaps, detergents, disinfectants, adhesives, plasticizers, rubber emulsifiers, corrosion inhibitors, and mining flotation reagents.

UREA

Urea (H_2NCONH_2, carbamide; melting point: 135°C, density: 1.3230) is a colorless crystalline solid, somewhat hygroscopic, that sublimes unchanged under vacuum at its melting point and decomposes above the melting point at atmospheric pressure, producing ammonia (NH_3), isocyanic acid (HNCO), cyanuric acid [(HNCO)$_3$], biuret ($H_2NHCONHCONH_2$), and several other minor products. Urea is very soluble in water (being a component of urine), soluble in alcohol, and slightly soluble in ether.

There are several approaches to the manufacture of urea, but the principal method is that of combining carbon dioxide with ammonia to form ammonium carbamate (Figs. 1 and 2):

$$CO_2 + 2NH_3 \rightarrow NH_2COONH_4$$

This exothermic reaction is followed by an endothermic decomposition of the ammonium carbamate:

$$NH_2COONH_4 \rightarrow NH_2CONH_2 + H_2O$$

Both are equilibrium reactions. The formation reaction goes to virtual completion under usual reaction conditions, but the decomposition reaction is less complete. Unconverted carbon dioxide and ammonia, along with undecomposed carbamate, must be recovered and reused.

In the process, a 2:1 molar ratio of ammonia and carbon dioxide (excess ammonia) are heated in the reactor for 2 hours at 190°C and 1500 to 3000 psi (10.3 to 20.6 MPa)to form ammonium carbamate, with most of the heat of reaction carried away as useful process steam. The carbamate decomposition reaction is both slow and endothermic. The mix of unreacted reagents and carbamate flows to the reactor-decomposer. The reactor must be heated to force the reaction to proceed. For all the unreacted gases and undecomposed carbamate to be removed from the product, the urea must be heated at lower pressure (400 kPa). The reagents are reacted and

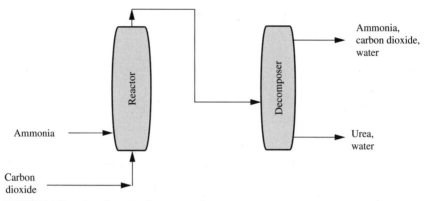

FIGURE 1   Once-through process for urea manufacture.

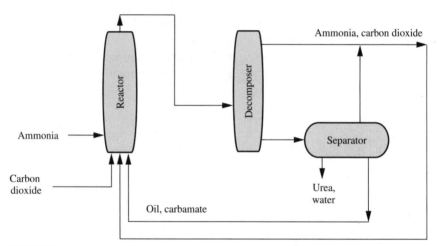

FIGURE 2   Recycle process for urea manufacture.

pumped back into the system. Evaporation and prilling or granulating produce the final product.

The mixture formed is approximately 35% urea, 8% ammonium carbamate, 10% water, and 47% ammonia. It is cooled to 150°C and the ammonia is distilled at 60°C. The residue from the ammonia still enters the crystallizer vessel at 15°C. More ammonia is removed by vacuum. The resulting slurry is centrifuged. All excess nitrogenous materials are combined and processed into liquid fertilizer, which contains a mixture of all these materials.

The corrosive nature of the reactants usually requires the reaction vessels to be lined with lead, titanium, zirconium, silver, or stainless steel. In the second step of the process, only about one-half of the ammonium carbamate is dehydrated in the first pass. Thus, the excess carbamate, after separation from the urea, must be recycled to the urea reactor or used for other products, such as the production of ammonium sulfate $[(NH_4)_2SO_4]$.

Urea is used as a solid fertilizer, a liquid fertilizer and miscellaneous applications such as animal feed, urea, formaldehyde resins, melamine, and adhesives. Presently, the most popular nitrogen fertilizer is a urea-ammonium nitrate solution. Urea-formaldehyde resins have large use as a plywood adhesive. Melamine-formaldehyde resins are used as dinnerware and for extra-hard surfaces (Formica®). The melamine is synthesized by condensation of urea molecules.

As a fertilizer, urea is a convenient form for fixed nitrogen and has the highest nitrogen content (46% by weight) available in a solid fertilizer. It is easy to produce as prills or granules and easily transported in bulk or bags with no explosive hazard. It dissolves readily in water and leaves no salt residue after use on crops and can often be used for foliar feeding.

Urea is also used as a protein food supplement for ruminants, in melamine production, and as an ingredient in the manufacture of pharmaceuticals (e.g., barbiturates), synthetic resins, plastics (urethanes), adhesives, coatings, textile antishrink agents, and ion exchange resins. It is an intermediate in the manufacture of ammonium sulfamate, sulfamic acid, and pthalocyanines.

UREA RESINS

Urea resins (urea formaldehyde polymers) are formed by the reaction of urea with formaldehyde (Fig. 1). Monomethylolurea ($HOH_2CNHCONH_2$) and dimethylolurea (($HOH_2CNHCONHCH_2OH$) are formed first under alkaline conditions. Continued reaction under acidic conditions gives a fairly linear, low-molecular-weight intermediate polymer.

A catalyst and controlled temperature are also needed and, since the amine may not be readily soluble in water or formalin at room temperature, it is necessary to heat it to about 80°C to obtain the methylol compounds for many amine-formaldehyde resins.

Heating for an extended period of time under acidic conditions will give a complex thermoset polymer of poorly defined structure including ring formation.

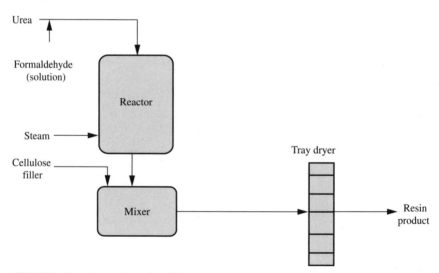

FIGURE 1   Manufacture of urea-formaldehyde resins.

VALIUM

Valium (diazepam) is a substituted benzodiazepine made by a series of reactions, one of which involves cyclization.

It is prepared by treating *p*-chloromethylaniline with benzoyl chloride and hydroxylamine to produce the benzophenone oxime. Reaction of the oxime with chloroacetyl chloride in the presence of sodium hydroxide, and subsequent reduction, yields diazepam.

*See **Benzodiazepines.***

VINYL ACETATE

The older process for the production of vinyl acetate (melting point: –93.2°C, boiling point: 72.3°C, density: 0.9317) involved the reaction of acetylene with acetic acid in the liquid phase with zinc amalgam as the catalyst.

$$CH \equiv CH + CH_3CO_2H \rightarrow CH_2=CHOCOCH_3$$

A newer method is based on the reaction of acetic acid with ethylene and has replaced the older acetylene chemistry.

$$2CH_2=CH_2 + 2CH_3CO_2H + O_2 \rightarrow 2CH_2=CHOCOCH_3$$

A Wacker catalyst is used in this process, similar to that for the manufacture of acetic acid. Since the acetic acid can also be made from ethylene, the basic raw material is solely ethylene. A liquid-phase process has been replaced by a vapor-phase reaction run at 70 to 140 psi and 175 to 200°C. Catalysts may be (1) carbon–palladium chloride–cupric chloride (C-$PdCl_2$-$CuCl_2$), (2) palladium chloride–alumina ($PdCl_2$-Al_2O_3), or (3) palladium–carbon–potassium acetate (Pd-C-KOAc). The product is distilled into water, acetaldehyde that can be recycled to acetic acid, and the pure colorless liquid, which is collected at 72°C. The yield is 95percent.

The reaction is conducted in a fixed-bed tubular reactor and is highly exothermic. With proper conditions, the only significant by-product is carbon dioxide. Enough heat is recovered as steam to perform the recovery distillation. Reaction is at 175 to 200°C under a pressure of 475 to 1000 kPa. To prevent polymerization, an inhibitor such as diphenylamine or hydroquinone is added.

Vinyl acetate is used for the manufacture of poly(vinyl acetate) resins (Fig. 1), poly(vinyl alcohol), and poly(vinyl butyral). Poly(vinyl acetate) is used primarily in adhesives, coatings, and paints. Copolymers of poly (vinyl acetate) with poly(vinyl chloride) are used in flooring, phonograph records, and PVC pipe. Poly(vinyl alcohol) is used in textile sizing, adhesives, emulsifiers, and paper coatings. Poly(vinyl butyral) is the plastic inner liner of most safety glass.

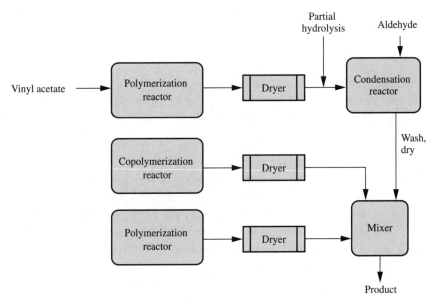

FIGURE 1   Manufacture of polyvinyl acetate resins.

VINYL CHLORIDE

Vinyl chloride (1-choroethylene; boiling point –31.6°C, density: 1.2137, flash point: –16°C) is manufactured by the addition of hydrogen chloride to acetylene in the presence of mercuric (Hg^{2+}) salts.

$$HC{\equiv}CH + HCl \rightarrow CH_2{=}CHCl$$

The process is combined with the process in which hydrogen chloride is produced by thermal dehydrochlorination of ethylene dichloride.

Thus, vinyl chloride is manufactured by the thermal dehydrochlorination of ethylene dichloride at 95 percent yield at temperatures of 480 to 510°C under a pressure of 50 psi with a charcoal catalyst.

$$CH_2ClCH_2Cl \rightarrow CH_2{=}CHCl + HCl$$

Vinyl chloride is separated from ethylene dichloride by fractional distillation. Although the conversion is low, 50 to 60 percent, recycling the ethylene dichloride allows an overall 99 percent yield.

More modern processes use the oxychlorination concept (Fig. 1) in which the vinyl chloride is produced from ethylene, chlorine, and oxygen.

$$2CH_2{=}CH_2 + HCl + O_2 \rightarrow 2CH_2ClCH_2Cl + 2H_2O$$

$$2CH_2ClCH_2Cl \rightarrow CH_2{=}CHCl + HCl$$

Vinyl chloride readily polymerizes, so it is stabilized with inhibitors to polymerization during storage.

The single use of vinyl chloride is in the manufacture of poly(vinyl chloride) plastic, which finds diverse applications in the building and construction industry as well as in the electrical, apparel, and packaging industries. Poly(vinyl chloride) does degrade relatively fast for a polymer, but various heat, ozone, and ultraviolet stabilizers make it a useful polymer. A wide variety of desirable properties can be obtained by using various amounts of plasticizers, such that both rigid and plasticized poly(vinyl chloride) have large markets. A lesser amount of the produced vinyl chloride is used for production of chlorinated solvents.

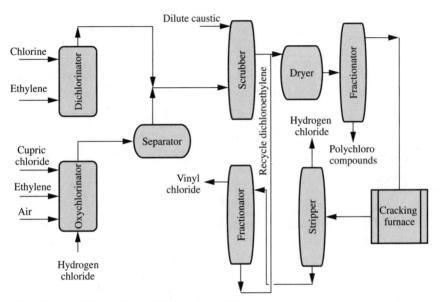

FIGURE 1   Manufacture of vinyl chloride by the oxychlorination process.

VINYL ESTERS

The addition of acids to acetylene furnishes the respective esters by addition across the double bond.

$$CH{\equiv}CH + CH_3COOH \rightarrow CH_3COOCH{=}CH_2$$

$$CH{\equiv}CH + HCl \rightarrow CH_2{=}CHCl$$

If two molecules of acid react, a compound such as 1,1-ethane diacetate is formed:

$$CH{\equiv}CH + 2CH_3COOH \rightarrow (CH_3COO)_2CHCH_3$$

Vinyl chloride is usually prepared by the oxychlorination (dehydrochlorination) of ethylene.

$$CH_2{=}CH_2 + Cl_2 \rightarrow CH_2ClCH_2Cl$$

$$CH_2{=}CH_2 + O_2 + 2HCl \rightarrow CH_2ClCH_2Cl + H_2O$$

$$CH_2ClCH_2Cl \rightarrow CH_2{=}CHCl + HCl$$

The cupric chloride ($CuCl_2$) catalyst (on an inert fixed carrier) may react as follows:

$$CH_2{=}CH_2 + 2CuCl_2 \rightarrow CH_2ClCH_2Cl + Cu_2Cl_2$$

$$2Cu_2Cl_2 + O2 \rightarrow CuO\ CuCl_2$$

$$CuOCuCl_2 + 2HCl \rightarrow 2CuCl_2 + H_2O$$

Exposure to vinyl chloride vapors, even in very small concentrations, causes some workers to develop liver cancer. The government requires that worker exposure to vinyl chloride monomer be no more than 1 ppm over an 8-hour period, and no more than 5 ppm for any 15-minute period. To achieve this requires extensive and expensive pollution-abatement systems.

Vinyl acetate is also made from ethylene in a vapor-phase process. The feed mixture is ethylene, acetic acid, and oxygen, and is circulated through a fixed-bed tubular reactor. The catalyst is a noble metal, probably palladium, and has a life of several years.

See also **Vinyl Acetate, Vinyl Chloride, Vinyl Ethers,** and **Vinyl Fluoride.**

VINYL ETHERS

The principal commercial vinyl ethers are methyl vinyl ether (methoxyethene, boiling point: 5.5°C, density: 0.7311, flash point –56°C); ethyl vinyl ether (ethoxyethene, boiling point: 35.7°C, density: 0.7541, flash point: –18°C); and butyl vinyl ether (1-ethenyloxybutane, boiling point: 93.5°C, density: 0.7792, flash point: –1°C). Others such as the *iso*-propyl, *iso*-butyl, hydroxybutyl, decyl, hexadecyl, and octadecyl ethers, as well as the divinyl ethers of butanediol vinyl ethers are miscible with nearly all organic solvents.

The principal methods of manufacture of vinyl ethers utilize vinylation of alcohols or cracking of acetals.

Vinyl ethers undergo all of the expected reactions of olefinic compounds plus a number of other reactions. For example, vinyl ethers react with alcohols give acetals. The acetals are stable under neutral or alkaline conditions and are easily hydrolyzed with dilute acid after other desired reactions have occurred. Reaction of a vinyl ether with water gives acetaldehyde and the corresponding alcohol and reaction of vinyl ethers with carboxylic acids gives 1-alkoxyethyl esters and with thiols gives thioacetals.

Hydrogen halides react vigorously with vinyl ethers to give 1-haloethyl ethers, which are reactive intermediates for further synthesis. Conditions must be carefully selected to avoid polymerization of the vinyl ether, Hydrogen cyanide adds at high temperature to give a 2-alkoxypropionitrile.

Chlorine and bromine add vigorously, giving, with proper control, high yields of 1,2-dihaloethyl ethers. In the presence of an alcohol, halogens add as hypohalites, which give 2-haloacetals. With methanol and iodine, this is used as a method of quantitative analysis by titrating unconsumed iodine with standard thiosulfate solution.

Vinyl ethers serve as a source of vinyl groups for transvinylation of such compounds as 2-pyrrolidinone or caprolactam.

VINYL FLUORIDE

Vinyl fluoride (CH_2=CHF; melting point: $-160.5°C$, boiling point: $-72.2°C$), the monomer for poly(vinyl fluoride), is manufactured by addition of hydrogen fluoride to acetylene:

$$HC \equiv CH + HF \rightarrow CH_2{=}CHF$$

VINYLIDENE CHLORIDE

Vinylidene chloride can be prepared by the reaction of 1,1,2-trichloroethane (prepared by the chlorination of vinyl chloride) with aqueous alkali.

$$CH_2=CHCl + Cl_2 \rightarrow CH_2ClCHCl_2$$

$$2CH_2ClCHCl_2 + Ca(OH)_2 \rightarrow 2CH_2=CCl_2 + CaCl_2 + 2H_2O$$

Other methods are based on bromochloroethane, trichloroethyl acetate, tetrachloroethane, and catalytic cracking of trichloroethane. Catalytic processes produce hydrogen chloride as a by-product.

The most common commercial process for the manufacture of vinylidene chloride is the dehydrochlorination of 1,1,2-trichloroethane with lime or caustic in slight excess (2 to 10%). A continuous liquid-phase reaction at 98 to 99°C gives a 90 percent yield of vinylidene.

Washing with water, drying, and fractional distillation purifies vinylidene chloride. It forms an azeotrope with 6% by weight of methyl alcohol, and purification can be achieved by distillation of the azeotrope, followed by extraction of the methanol with water; an inhibitor is usually added at this point.

Commercial grades of vinylidene fluoride may contain an inhibitor such as the monomethyl ether of hydroquinone (MEHQ). This inhibitor can be removed by distillation or by washing with 25% by weight aqueous caustic under an inert atmosphere at low temperatures.

VINYLIDENE FLUORIDE

Vinylidene fluoride is manufactured by the thermal elimination of hydrogen chloride from 1-chloro-1,1-difluoroethane. The starting material (1-chloro-1,1-difluoroethane) is manufactured by any of several different routes.

$$CH\equiv CH + 2HF \rightarrow CH_3CHF_2$$

$$CH_3CHF_2 + Cl_2 \rightarrow CH_3CClF_2 + HCl$$

$$CH_2=CCl_2 + 2HF \rightarrow CH_3CClF_2 + HCl$$

$$CH_3CCl_3 + 2HF \rightarrow CH_3CClF_2 + 2HCl$$

$$CH_3CClF_2 \rightarrow CH_2=CF_2$$

Dehydrohalogenation of 1-bromo-1,1-difluoroethane or 1,1,1-trifluoroethane, or dehalogenation of 1,2-dichloro-1,1-difluoroethane are alternative routes.

1-chloro-1,1-difluoroethane can also be continuously prepared by the pyrolysis of trifluoromethane (CHF_3) in the presence of a catalyst and either methane or ethylene.

WATER GAS

Water gas is often called *blue gas* because of the color of the flame when it is burned. It is produced by the reaction of steam on incandescent coal or coke at temperatures above 1000°C.

$$C + H_2O \rightarrow CO + H_2$$
$$C + 2H_2O \rightarrow CO_2 + 2H_2$$

The heating value of this gas is low (<300 Btu/ft^3) and, to enhance it, oil may be atomized into the hot gas to produce *carbureted* water gas, which has a higher heat content.

WAX

Paraffin wax is a solid crystalline mixture of straight-chain (normal) hydrocarbons ranging from C_{20} to C_{30} and possibly higher, that is, $CH_3(CH_2)_nCH_3$ where $n \geq 18$. It is distinguished by its solid state at ordinary temperatures [25°C (77°F)] and low viscosity [35 to 45 SUS at 99°C (210°F)] when melted. However, in contrast to petroleum wax, petrolatum (*petroleum jelly*), paraffin wax does in fact contain both solid and liquid hydrocarbons. It is essentially a low-melting, ductile, microcrystalline wax.

Paraffin wax from a solvent dewaxing operation is commonly known as *slack wax*, and the processes employed for the production of waxes are aimed at deoiling the slack wax (petroleum wax concentrate).

Wax sweating was originally used to separate wax fractions with various melting points from the wax obtained from shale oils and is still used to some extent, but is being replaced by the more convenient wax recrystallization process. In wax sweating, a cake of slack wax is slowly warmed to a temperature at which the oil in the wax and the lower melting waxes become fluid and drip (or sweat) from the bottom of the cake, leaving a residue of higher-melting wax.

Wax recrystallization, like wax sweating, separates slack wax into fractions, but instead of using the differences in melting points, it makes use of the different solubility of the wax fractions in a solvent, such as the ketone used in the dewaxing process. When a mixture of ketone and slack wax is heated, the slack wax usually dissolves completely, and if the solution is cooled slowly, a temperature is reached at which a crop of wax crystals is formed. These crystals will all be of the same melting point, and if they are removed by filtration, a wax fraction with a specific melting point is obtained. If the clear filtrate is further cooled, a second crop of wax crystals with a lower melting point is obtained. Thus, by alternate cooling and filtration, the slack wax can be subdivided into a large number of wax fractions, each with different melting points.

The melting point of paraffin wax has both direct and indirect significance in most wax utilization. All wax grades are commercially indicated in a range of melting temperatures rather than at a single value, and a range of 1°C (2°F) usually indicates a good degree of refinement. Other common physical properties that help to illustrate the degree of refinement of the wax are color, oil content, API gravity, flash point, and viscosity, although the last three properties are not usually given by the producer unless specifically requested.

Petroleum waxes (and petrolatum) find many uses in pharmaceuticals, cosmetics, paper manufacturing, candle making, electrical goods, rubber compounding, textiles, and many more.

WOOD CHEMICALS

Wood chemicals are derived from woody plants made of strong, relatively thick-walled long cells that make good fibers. The cell wall in this type of plant is a complex mixture of polymers that varies in composition. The solid portion of wood is over 95% organic material that is a mixture of three groups of polymers:

1. Cellulose, which is approximately 45% of the dry weight in an ordered array of high-molecular-weight glucose polymer chains, currently most valuable as fiber

2. Hemicellulose (20 to 25%), which is a disordered array of several sugar polymers for which there is currently no economical use except as fuel

3. Lignin (20 to 25%), which serves as binder for the cellulose fibers, and is a complex amorphous polyphenol polymer

Wood also contains extractives—organics removable with inert solvents—and the extracts vary with the species and the location in the tree.

Lignin has been described as *the adhesive material of wood* because it cements the fibers together for strength. It is a complex cross-linked polymer of condensed phenylpropane units joined together by various ether and carbon linkages. Lignin can be considered to be a polymer of coniferyl alcohol. About 50 percent of the linkages are β-aryl ethers. Lignin can be degraded with strong alkali, with an acid sulfite solution, and with various oxidizing agents. It is therefore removed from the wood to leave cellulose fibers, commonly called *pulp*. Although there are many differences between hardwood and softwood, the hardwoods always have less lignin and more hemicellulose (high in xylose), whereas the softwoods have more lignin and less hemicellulose (which is high in galactose, glucose, and mannose units). Besides the holocellulose and lignin of the cell wall, wood contains about 25% by weight of material that can be recovered by steam distillation or by solvent extraction.

In the kraft process (Fig. 1), the focus is a closed system with the exception of sodium sulfate being added periodically. Only wood enters the loop

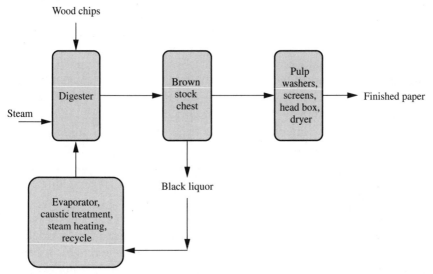

FIGURE 1   The kraft process.

and only pulp leaves. Digestion of the wood–white liquor mixture occurs at 170 to 175°C and 100 to 135 psi for 2 to 5 hours. A typical digester is 40 ft high with a diameter of 20 ft and can hold up to 35 tons of wood chips at a 1:4 wood: white liquor ratio.

The resulting pulp is separated from the black liquor (colored with organics), which is then oxidized to sodium thiosulfate ($Na_2S_2O_3$) and further oxidized in the furnace to sodium sulfate (Na_2SO_4). The organic material from the digestion process is oxidized in the furnace to carbon dioxide whereas the sodium sulfate is reduced back to sodium sulfide (Na_2S), the original oxidation state of sulfur in the process. The carbon dioxide is absorbed by sodium hydroxide (NaOH) to form sodium carbonate (Na_2CO_3). Water is added to the material from the furnace, forming a green liquor containing sodium hydrosulfide (NaSH) and sodium hydroxide. The sodium carbonate is reacted with calcium oxide (CaO) and water to give more sodium hydroxide (causticizing) and calcium carbonate ($CaCO_3$) which is usually filtered and transformed on site back into calcium oxide by a lime kiln.

After the crude pulp is obtained from the alkaline sulfate process, it is bleached in stages with elemental chlorine, extracted with sodium hydroxide, and oxidized with calcium hypochlorite, chlorine dioxide, and hydrogen peroxide. This lightens it from a brown to a light brown or even white (difficult) color. Chlorination of the aromatic rings of residual lignin is probably what is occurring, although this has not been completely studied. Typical end uses of kraft pulp are brown bags, paper boxes, and milk cartons.

Much of the methyl mercaptan and dimethyl sulfide can be oxidized to dimethylsulfoxide, a useful side product that is a common polar, aprotic solvent in the chemical industry. This is in fact the primary method of its manufacture, as a kraft by-product. Caution must be used when handling it because of its extremely high rate of skin penetration.

Two other important side products of the kraft process are *turpentine* and *tall oil*. The turpentine is obtained from the gases formed in the digestion process. Tall oil soap is a black viscous liquid of rosin and fatty acids that can be separated from the black liquor by centrifuging. Acidification gives tall oil.

The *acid sulfite process* is used to obtain a higher-quality paper. It is also more water polluting. Digestion occurs in a mixture of sulfur dioxide and calcium or magnesium bisulfite. The magnesium bisulfite process is better for pollution but still not so good as the kraft process. Sulfite pulp is used for bond paper and high-grade book paper.

Wood is the source of a large number of chemicals and pharmaceuticals, and a number of lower-volume chemicals can be obtained from wood hydrolysis (Fig. 2). Furfural is formed from the hydrolysis of some polysaccharides to pentoses, followed by dehydration.

Furfural is used in small amounts in some phenol plastics; it is a small minor pesticide and an important commercial solvent. It can be converted into the common solvent tetrahydrofuran and an important solvent and intermediate in organic synthesis, furfuryl alcohol.

Vanillin is obtained from sulfite waste liquor by further alkaline hydrolysis of lignin. It is the same substance that can be obtained from vanilla bean extract and is the common flavoring in foods and drinks. Natural and synthetic vanillin can be distinguished from each other by a slight difference in the amount of ^{13}C in their structure, since one is biosynthetic in the bean and the other is isolated from a second natural product, wood, by hydrolysis of the lignin.

Charcoal was a valued commodity in antiquity. The ancient Egyptians used the volatile product of hardwood distillation, pyroligneous acid, for embalming. Before synthetic organic chemistry became well established, destructive hardwood distillation provided several important industrial chemicals, among these were acetone, acetic acid, and methanol (still often referred to as wood alcohol). Charcoal is a fine, smokeless fuel, prized for its smokeless nature and used extensively for outdoor cooking. Acetone was originally made by the dry distillation of calcium acetate made from wood-derived acetic acid, but better, cheaper sources are also available.

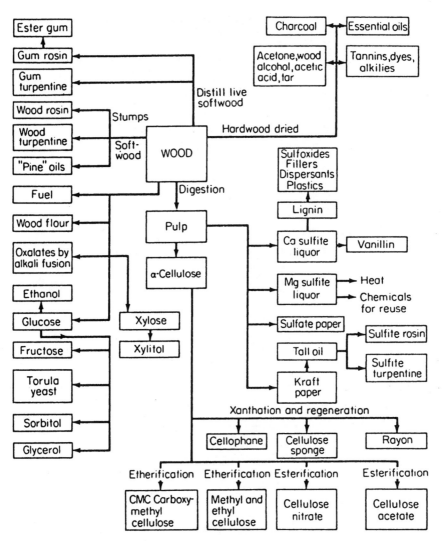

FIGURE 2   Chemicals from wood.

The manufacture of charcoal, especially briquettes, has been increasing. It is the residue after combustion of the volatiles from a hardwood distillation. It consists of elemental carbon and incompletely decomposed organic material and many adsorbed chemicals. Carbonization is usually performed at about 400 to 500°C. The charcoal has a volatile content of 15 to 25% by weight and can be made in about 37 to 46 percent yield by weight from wood.

*See **Pulp and Paper Chemicals**, **Tall Oil**, **Turpentine**.*

XENON

*See **Rare Gases**.*

XYLENES

The general term *xylenes* (*xylols*) refers to the C_8 aromatic isomers $(CH_3C_6H_4CH_3)$ with the methyl groups in positions ortho, meta, and para to each other. Thus, the xylenes consist of three isomers: *o*-xylene, *m*-xylene, and *p*-xylene

The xylenes are colorless, flammable liquids with properties that are significantly different from each other or significantly similar to each other, depending upon the perspective and the need for separation of the isomers.

	Melting point	Boiling point	Density
o-xylene	–25°C	144.0°C	0.8968
m-xylene	–47.4°C	139.1°C	0.8684
p-xylene	13.2°C	138.5°C	0.8611

o-xylene (flash point: 27°C) can be separated from the *meta*-isomer (flash point: 27°C) and the *para*-isomer (flash point: 32°C) by distillation; the *para*-isomer and the *meta*-isomer are difficult to separate by distillation. The xylenes are insoluble in water, soluble in alcohol and ether.

The term *mixed xylenes* describes a mixture containing the three xylene isomers. Commercial sources of mixed xylenes include catalytic reformate, pyrolysis gasoline, toluene disproportionation product, and coke-oven light oil. Ethylbenzene is present in all of these sources except toluene disproportionation product.

In catalytic reforming, a low octane naphtha cut (typically a straight run or hydrocracked naphtha) is converted into high-octane aromatics, including benzene, toluene, and mixed xylenes. Aromatics are separated from the reformate by using a solvent such as diethylene glycol or sulfolane and then stripped from the solvent. Distillation is then used to separate the benzene-toluene-xylene into its components.

The amount of xylenes contained in the catalytic reformate depends on the fraction and type of crude oil, the reformer operating conditions, and the catalyst used. The amount of xylenes produced can vary widely, typically ranging from 18 to 33% by volume of the reformate. In the United States, only about 12% of the xylenes produced via catalytic reforming are actually recovered for use as petrochemicals. The unrecovered reformate xylenes are used in the gasoline pool.

Pyrolysis gasoline is a by-product of the steam cracking of hydrocarbon feeds in ethylene crackers. Pyrolysis gasoline typically contains about 50 to 70% by weight of aromatics, of which roughly 50% is benzene, 30% is toluene, and 20% is mixed xylenes (which includes ethylbenzene).

Coke oven light oil is a by-product of the manufacture of coke for the steel industry. When coal is subjected to high-temperature carbonization, it yields 16 to 25 liters/ton of light oil that contains 3 to 6% by volume of mixed xylenes.

Xylenes can also be manufactured by the disproportionation of toluene (Fig. 1):

$$2C_6H_5CH_3 + \ \rightarrow \ C_6H_6 + C_6H_4(CH_3)_2$$

or by the transalkylation of toluene with trimethylbenzenes.

$$C_6H_5CH_3 + C_6H_3(CH_3)_3 \ \rightarrow \ 2C_6H_4(CH_3)_2$$

Toluene disproportionation is a catalytic process in which 2 moles of toluene are converted to 1 mole of xylene and 1 mole of benzene. Although the mixed xylenes from toluene disproportionation are generally more costly to produce than those from catalytic reformate or pyrolysis gasoline,

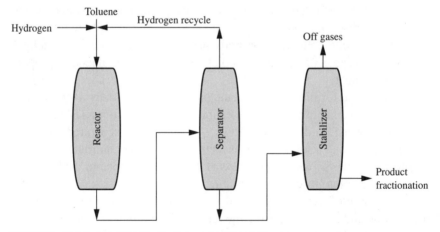

FIGURE 1 Xylene manufacture by the disproportionation of toluene.

their principal advantage is that they are very pure and contain essentially no ethylbenzene.

The manufacture of mixed xylenes and the subsequent production of high-purity o-xylene and p-xylene consists of a series of stages in which (1) the mixed xylenes are initially produced, (2) o-xylene and p-xylene are separated from the mixed xylenes stream, and (3) the p-xylene–depleted xylene stream is isomerized back to an equilibrium mixture of xylenes and then recycled back to the separation step.

The two principal methods for producing xylenes are catalytic reforming and toluene disproportionation.

In one process (Fig. 2), a light fraction (boiling range 65 to 175°C) from a straight run petroleum fraction or from an isocracker is fed to a catalytic reformer and is followed by fractionation and extraction. The mixed xylenes stream is then processed further to produce high-purity p-xylene and/or o-xylene. Because of the close boiling points of p-xylene and m-xylene, production of high-purity p-xylene by distillation is impractical and methods such as crystallization and adsorption are used.

In the separation of the xylenes, the C_8 mixture is cooled to –70°C in the heat exchanger refrigerated by ethylene. Because of the difference in melt-

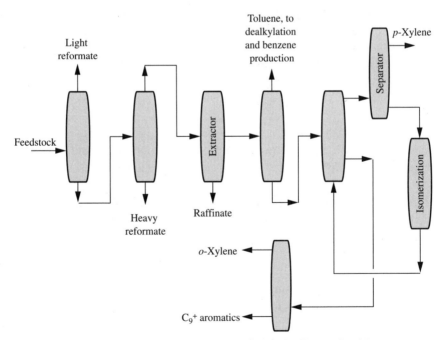

FIGURE 2 Manufacture of xylenes by catalytic reforming and toluene disproportionation.

ing points (*o*-xylene –25.0°C; *m*-xylene – 47.9°C; *p*-xylene 13.2°C), the para-isomer crystallizes preferentially. The other two isomers remain liquid as a mixture and the solid para-isomer is centrifuged and separated. A second cooling cycle needs only propane as coolant, and complete separation is accomplished with an optional third cooling cycle.

Because of the large demand for *p*-xylene, another method is now being used to increase the percentage of the para-isomer in mixed xylenes. They are heated at 300°C with an acidic zeolite catalyst, which equilibrates the three xylenes to an *o*, *m*, *p* ratio of 10:72:18. The para-isomer is separated by fractional crystallization, whereas the ortho-meta mixture is reisomerized with the catalyst to produce more para-isomer.

Another possibility for separating the para-isomer involves selective adsorption on zeolites, then desorption after the ortho and meta isomers have been separated. The slightly different boiling point of the *o*-xylene is the basis for separation from the other two isomers through an elaborate column.

After separation of the preferred xylenes, i.e., *p*-xylene or *o*-xylene, the remaining raffinate stream, which tends to be rich in *m*-xylene, is typically fed to a xylenes isomerization unit in order to further produce the preferred xylenes. Isomerization units are fixed-bed catalytic processes that are used to produce a close-to-equilibrium mixture of the xylenes. To prevent the buildup of ethylbenzene in the recycle loop, the catalysts are also designed to convert ethylbenzene to either benzene and xylenes or to benzene and diethylbenzene.

Historically, the isomerization catalysts have included amorphous silica-alumina, zeolites, and metal-loaded oxides. All of the catalysts contain acidity, which isomerizes the xylenes and if strong enough can also crack the ethylbenzene and xylenes to benzene and toluene.

The xylenes can be used as a mixture or separated into pure isomers, depending on the application. The mixture is obtained from catalytic reforming of naphtha (using a platinum catalyst) and separated from benzene and toluene by distillation.

p-xylene is used in the manufacture of terephthalic acid, which is reacted with ethylene glycol to give poly (ethylene terephthalate). Large amounts of this polyester are used in textile fibers, photographic film, and soft drink bottles.

o-xylene is used in the manufacture of phthalic anhydride, an intermediate in the synthesis of plasticizers, substances that make plastics more flexible. A common plasticizer is dioctyl phthalate.

*See **Benzene** and **Ethylbenzene**.*

ZINC CHROMATE

Zinc chromate ($ZnCrO_4$), also known as zinc yellow (with the approximate formula $4ZnO \cdot K_2O \cdot 4CrO_3 \cdot 3H_2O$), is used as a pigment because of its excellent corrosion-inhibiting effect both in mixed paints and as a priming coat for steel and aluminum.

ZINC OXIDE

Zinc oxide (ZnO) is manufactured by oxidizing zinc vapor in burners in which the concentration of zinc vapor and the flow of air are controlled to produce the desired particle size and shape. The hot gases and particulate oxide or fume pass through tubular coolers, and then the zinc oxide is separated in a baghouse. The purity of the zinc oxide depends upon the source of the zinc vapor.

In the *indirect process*, zinc metal vapor for burning is produced in several ways, one of which involves horizontal retorts. Since the entire vapor is burned in a combustion chamber, the purity of the oxide depends on that of the zinc feed. Oxide of the highest purity requires special high-grade zinc, and less-pure products are made by blending in Prime Western and even scrap zinc.

In the *direct process*, four or more firebrick furnaces having common walls are charged in cyclic fashion. Coal that is hot from the previous charge is first spread on the grate and, after ignition, a damp, well-blended mixture of zinc ore or zinc-containing material and coal is added. The bed is maintained in a reducing condition with carbon monoxide to produce zinc and lead, if present. Metal vapors are drawn into a chamber above the furnace, where combustion air oxidizes them to pigment. The hot pigment-gas stream enters a cooling duct common to the whole block and, in this way, the product becomes a uniform blend.

Traveling-grate furnaces can also be employed. In this process, anthracite briquettes are fed to a depth of about 15 cm. After ignition by the previous charge, the coal briquettes are covered by ore/coal briquettes. The latter are dried with waste heat from the furnace. Zinc vapor evolves and burns in a combustion chamber, and the spent clinker falls into containers for removal.

A pigment-grade zinc oxide rotary kiln uses high temperature to produce pigment-quality zinc oxide and makes possible higher recovery than a grate furnace.

Other processes include an *electrothermic process*, an electric-arc vaporizer process, and the slag fuming process.

Zinc oxide, as an amphoteric material, reacts with acids to form zinc salts and with strong alkali to form zincates. In the vulcanization of rubber, the chemical role of zinc oxide is complex and the free oxide is required, probably as an activator.

Zinc oxide reacts with organic acids to produce zinc soaps and also reacts with carbon dioxide in moist air to form oxycarbonate. Acidic gases, e.g., hydrogen sulfide, sulfur dioxide, and chlorine, react with zinc oxide, and carbon monoxide or hydrogen reduce it to the metal. At high temperatures, zinc oxide replaces sodium oxide in silicate glasses. An important biochemical property of the oxide is its fungicidal/mildewstatic action. It is also soluble in body fluids and soils.

Zinc oxide of high purity is required for pharmaceutical, photoconductive, and certain other uses, and is manufactured by the indirect process. Less-pure zinc oxide is manufactured by the direct process, by which impure zinc oxide is reduced to zinc vapor that is then burned.

ZINC SULFATE

*See **Lithopone.***

ZINC SULFIDE

*See **Lithopone**.*

INDEX

ABOUT THE AUTHOR

James G. Speight is the author/editor/compiler of more than 20 books and bibliographies related to fossil fuel processing and environmental issues. As a result of his work, Dr. Speight was awarded the Diploma of Honor, National Petroleum Engineering Society, for Outstanding Contributions in the Petroleum Industry in 1995 and the Gold Medal of Russian Academy of Natural Sciences for Outstanding Work. He was also awarded the Degree of Doctor of Science from the Russian Petroleum Research Institute in St. Petersburg.